INSECT REMEDIES
AND ONE SPIDER

Jenny Jones

Illustrations: Crispin Chetwynd

Cover: Joshua Bennet Chetwynd

Layout: Crispin Chetwynd

First Printing: 2012

Published by Academy of Light Ltd.
102 Chaplin Road,
London, NW2 5PR

ISBN-13: 9781904472049

CONTENTS

ACKNOWLEDGMENTS

AM BOB FE PETHNUSAU

(For all my relations)

To Dr Yubraj Sharma who introduced me to the beauty of homeopathic remedies. Ceridwen Gwynne James, my teacher who has helped my understanding of working with nature spirits in the 'old ways'. Howard Charing who helped me rediscover my spiritual connections and taught me so much about shamanism, spirituality and plant medicine and first introduced me to life in the Peruvian jungle which ultimately led to these remedies and to Rolando the shaman who taught me about the medicinal plants and the wildlife in the Amazon.

(Jenny Jones)

FOREWORD

Jenny was first and foremost, a shaman conversant with the language of stones, crystal, the earth, the seasons, the animal and plant kingdoms. Inherent in her healing work, she utilised the spiritual forces in nature around her. She taught me what it means to be a shamanic healer through her practice and integrity. I am glad she was part of my physical reality, before she returned to higher dimensions in her ongoing soul's journey, and I know the same is felt by all her friends and family.

Jenny especially developed an organisational capacity - through provings, to capture the essence of the new homoeopathic remedies, and this selection of insect remedies is but a small part of a larger body of unfinished provings work. Like any new homoeopathic remedy, they can initially seem too esoteric or distant, until one realises patients are walking through the door who, in displaying symptoms in similimum, need these remedies.

Great homoeopathic remedies certainly do develop over time into richer constitutional features, to become useful polycrests, and I think Jenny has certainly birthed a few powerful remedies here.

So I hope you enjoy, through her words, Jenny's thoughts and visions, including any she continues to project in Spirit.

Dr. Yubraj Sharma, School of Shamanic Homoeopathy, London (2012)

INTRODUCTION

The 20 homoeopathic remedies described in this book can all be ordered from select homoeopathic pharmacies, as medicating remedies to drop onto sac lac etc, or as pre-medicated pillules. Their contact details are listed below. The School of Shamanic Homoeopathy and Academy of Light Ltd publisher wish for these remedies to be available to the wider homoeopathic community, in keeping with the last wishes of Jenny Jones, the originator of the whole project. We hope you are able to utilise the information in this book to expand on the materia medica profile of the remedies with provings confirmations, further provings symptoms or clinical confirmations. If there is any interesting feedback from this, do let us know.

Ainsworths Homoeopathic Pharmacy
36 New Cavendish Street, London W1G 8UF
Tel. +44 (0)20 7935 5330
london@ainsworths.com

Helios Homoeopathy Ltd
89-97 Camden Rd, Tunbridge Wells, Kent TN1 2QR England
Tel. +44 (0)1892 537254
pharmacy@helios.co.uk

FINDING THE INSECT REMEDIES

AYAHUASCA CEREMONIES

I took part in 'ayahuasca' ceremonies to help get in touch with the spirits of the herbs. These ceremonies were very powerful and put me in touch with many aspects of myself, my being, my understanding of how the spiritual world works and how the human dream world works, as well a helping me refine specific knowledge about herbs.

A lot is written about ayahuasca visions, so I am not going to go into to much detail here except to give you a bit of an insight into my experiences.

Ayahuasca is a herbal mix of basically two ingredients:
The vine Ayahuasca itself (Banisteriopsis Caapi) and Chakruna leaves (Psychoptria viridans).

These are chopped up, and boiled in water for several hours. Different shamen sometimes add other ingredients to help the ceremony to have a particular quality eg. Psilocybin (magic mushrooms) to help connect to earth spirits, or coca leaves, which help give a boost to the mental process and quality of thinking in the days after the ceremony. Ayahuasca is a hallucinogenic and a medicine plant. As such it can be misused, and I feel very strongly that anyone wanting to try ayahuasca should only do so in ceremony with a suitably competent trained shaman skilled in the use of Ayahuasca. Unfortunately there are many false shaman providing Ayahuasca ceremonies without this training. The result can result in people getting stuck in the experiences, and these can be more destabilising than those with other medicine plants.

Ayahuasca ceremonies begin when it is dark. Typically they are held in a hut used for that purpose. People gather and sit or lie on benches. The shaman comes in and by candlelight prepares the Ayahuasca and often sings beautiful icaros to it to open the ceremony and dedicate it.

People are then given the Ayahuasca to drink – it has an awful taste, some shamen give another drink afterwards to take the taste away, most do not.

Then, everything is silent – there is a quality of anticipation and expectation in the hut, and sometimes a feeling of building up energy in your body, but mostly it is quiet. Then suddenly either you or someone else vomits or need to go to the toilet (at this point most people start to move either around the hut or go outside).

The vomiting and purging is a cleaning process, which helps your body physically let go of any energies which aren't needed anymore. Some people are 'clear' and don't vomit every time – others purge a lot, sometimes from every orifice!

During the purging process the energy in my body often used to have a feeling of accumulating electrical charge until I started to purge. At that point, there would be a psychedelic quality to my experience, which seemed to relate to what I've read about LSD. You get very bright colours and lights, often if you are being sick the vomit itself looks like something else – jewels, sweets, insects etc. Sometimes you 'hear' funky or fairground music and the world has the quality of a cartoon.

As the 'purging' comes to an end, people generally are back in the hut and the shaman starts to sing an 'icaro'. Icaros are specific songs to the Ayahuasca, which help aid the visions but also are specific to help people in certain situations, e.g. Someone might have a vision where they get frightened and may get 'stuck' at that point, the shaman will then sense this and sing an appropriate icaro to help him move on.

Lying or sitting in the dark listening to the haunting singing of the icaros, the quality of the visions change from the psychedelic to visions which have a 3D quality – it is more like being in a three dimensional world than the 'normal' 2D quality of dreams.

At that point, I would set my intention for the visions – drink the herbal tincture and ask for connection to the spirit of the herb or ask for the

Ayahuasca to help me find the answer to a specific question I had. Surprising insights came. I had many and varied experiences, some of which were joyful, some of which were painful when I realised things about myself that I would have preferred not to look at. If I asked questions, I got answers and I did get a lot of information about the herbs both on a practical level and on an esoteric level.

During the visions, the shaman sings the icaros and also moves around singing to individuals sometimes fanning them and generally helping everyone make the most of the experience. After a time, usually a couple of hours, the visions start coming to an end and the shaman stops singing. Everyone starts to 'come back' to themselves and gradually they leave the ceremony and go back to their beds to sleep.

The following day has a different quality – a more dream like quality when you are very connected to your astral body and if you stay still and 'hang in a hammock' other dreams and insights come to you.

(The Ayahuasca visions the previous night initially had an electrical quality – connecting directly to the etheric body and sometimes helping to burn new neural pathways as a feature of the Ayahuasca experience is that the chemical messengers across the nervous synapses do not decay quickly as they do normally – which gives rise to the 3rd dimensional clearing from the etheric effects the energy and experiences move to the astral body.)

On the second day, after the ceremony, again if you are quiet and meditative the insights that come are more mental and logical and you can piece together all the information you have, in my case this was when I could collate the herbal knowledge with the insightful dreams from the 'spirit' of the plant, and either make a new homeopathic remedy or write down what I had learned.

So during the Ayahuasca experience there is a movement from the physical body purging, to the etheric body, astral body and mental body, all of which are involved in the experience of Walking between the Worlds and bringing a spiritual dimension to making sense of things in our illusion of the human world.

CONNECTING TO THE INSECTS

For several weeks, during our expeditions looking for herbs in the jungle, and during the Medicine Ceremonies, I had come to realise the importance of insects to the plant world, there was a huge connection, it was as if all the plants were directly connected to particular insects whilst other insects were there to help 'clean up' the detritus and help the plants grow and thrive without taking on old, obsolete energies. They transformed and transmuted old dying or dead things to make way for new life.

I got a great deal of joy from sitting in the 'Mirador', which was a beautiful viewing platform in the jungle overlooking a pond. There were always beautiful dragonflies and butterflies around, whilst in the water there were many larvae and other creatures, which often looked sinister but had a particular purpose. Seeing the transformation from ugly larvae with pincers and claws to beautiful dragonflies with transient ethereal bodies was an interesting parallel with how we spend most of our lives in the mundane human world of duality and materialism (the mud) and aspire to transform our lives and fly into the spiritual world where there is an interconnection between everything and a joy.

A friend on observing the insect life in the jungle said that 'everything is eating something else' and certainly if ever anything was left, thrown away or died it would be covered by insects within minutes. I had not considered making insect remedies, although in my practice at home I used apis and scorpion and some of the spider remedies.

Then one night, in an Ayahuasca ceremony, things changed. I had gone into the ceremony to work with the spirit of a herb and had taken some leaves in with me, which I intended to eat during the ceremony.

It was a beautiful night, one of those that was so clear and peaceful that you felt the myriad stars were so close to earth that you could reach up and touch them, and even though the moon was a thin crescent the starlight seemed to light the jungle clearing. It was as though the fairies and elves and nature spirits were just waiting to come out and play - there was a magical quality in the air.

I went into the ceremony and that night was not sick and didn't purge at all. I was transfixed by the stars and the magical atmosphere of the clearing, so didn't eat the herb, but fell into a vivid dream, I saw several insects in a space in the jungle – there were various beetles on the floor in patches of mud or soil, tree ants, a shiny green bee and a large blue and black butterfly. When the butterfly saw me it came over and flapped in front of my face at eye level and said that it wanted to be a homeopathic remedy. I explained that I was there to work with a herbalist to learn about existing jungle herbal remedies and that although I loved seeing the insects in the jungle I had no wish to kill any of them to turn them into remedies.

The butterfly – Ulysses, Ulysses – stayed still and became insistent and the other insects came closer and also came 'in my face' saying that they wanted to become remedies. They were very insistent, trying to get me to agree. However, I was resistant to this, partly because I had so much to learn about herbal medicine in the time I had left in the jungle and partly because I don't like to kill anything.

After what seemed like a long time but was probably only moments, we were at a standoff and I heard myself saying that if they really wanted to be remedies the insects would have to give themselves to me because I was not going to kill them. My vision disappeared and I went back to enjoying the magic of the starlight.

The ceremony had lasted a long time that night and I arrived back at my hut after midnight. There, on the top step, in front of the door was a dead blue butterfly the very one I had seen in the vision. That got my attention!

The next morning I walked to a tambo about a mile away from the village, which I spent a lot of time lying in a hammock meditating a collating my thoughts and information. As I arrived at the tambo again on the top step by the door, there was another dead blue butterfly. I realised that 'insect remedies' were part of what I needed to do.

A few days later I went into Iquitos to catch up with my email and one of the street children who I'd got to know quite well came up to me. (He used to teach me Spanish and would get anything I needed and I used to teach him English and feed him and help his family who were very poor and in ill health.)

He was carrying a paper bag and asked if I wanted it. It contained a lot of different jungle insects (including the green bee and some beetles I had seen in the vision). He explained that he usually sold them to a man who made them into pictures for tourists but the man was away so he didn't know what to do with them.

It would appear that the insects had arranged to be given to me. The next step was to make them into remedies.

Firstly, the insects were photographed – unfortunately I didn't have the right photographic equipment so some of these did not turn out very well. In some cases, people in Iquitos helped me identify the species from books they had there. I identified the rest when I came back to London using Libraries, the internet and the Natural History Museum.

Once the insect had been identified, I tried to find out as much about its lifestyle and correspondences as possible, as that gave a pointer to what the remedy could possibly be used for.

To make the remedies I used lactose powder and 96% or 98% alcohol. I stored the insects in alcohol before making the remedies. Generally I used a pestle and mortar and crushed the whole insect down in lactose powder and from the 'mother tincture' took a tiny amount which I added to 100 x more lactose powder, and crushed this together for some minutes. I repeated the process until I had the remedy in Lactose at 6c. Then I put some of the 6c lactose powder in a test tube of alcohol succuss (shake it) and left it for at least a day.

I took a drop of 98% alcohol and a drop from the alcohol with the dissolved lactose, mixed them together to get a 6c remedy in alcohol and used this to sucuss to 30c in alcohol – using 100 drops of alcohol to one drop of remedy for each dilution.

The information from the research about the insects, was also a starting point for the provings. However, at the School of Shamanic Homeopathy, we also did meditative provings and shamanic journeys to the spirits of the insects to get further information.

Other people 'proved' the remedies classically, taking them every day and seeing what 'symptoms' came up, keeping a diary log.

At some point the remedies were used with clients and this process of refining and defining these remedies is still ongoing.

When people knew I was making insect remedies, I was given other insects and some other arthropods and spiders some of which have also been made into remedies and are included here.

THE SIGNIFICANCE OF THE INSECTS AS HOMEOPATHIC REMEDIES

Insects are the most common forms of animal life on earth with over 800,000 species known and more being discovered all the time; yet in spite of this only a few classical homeopathic remedies are in common usage.

Little has been written about insect remedies as a class themselves. Peter Frazer in his book Insects-escaping the Earth (ISBN 9781874581 -18-5) makes some interesting observations as a starting point for study.

Esoterically, both Alice Bailey in her books and Helena Roerich in the Agni Yoga books have linked the preponderance of insects of the earth to people's thought processes. Basically, they are saying that most people have a lot of petty, irritating thoughts; thoughts of selfish intent coming from lower egos and that this chaotic thinking has precipitated down into physicality as the insects.

Where there is a lot of irritation or anger or resentment in the world this results in the physical irritations caused by biting insects and the noxious chemicals that can be carried by them. Thus the 'mentals' of the insect remedies are especially important. It has even been said that insects are the physical form of unwanted, discarded thoughts.

In anthroposophical medicine, much has been written about insects and the relationship with the cosmos e.g. bees carrying solar energy and also their interrelationship with plants, both generally as food shelter and pollination. In some cases where there is direct relationship between one insect and a particular plant – then often the plant and insect remedies can complement each other to work toward cure.

There is a very close relationship between the insects and the plants and mineral kingdoms. Some insects live in the earth whilst other can fly into the air element. Insects that live or pupate above ground bring an astrality into their lives. The plant stems and tree trunks are influenced by solar forces and grow vertically towards the light – they are 'thrust up the coil' reaching out to escape the earth and join with the air element.

> Some of the insects have a close relationship with plants like the solanaceae which have excess astral forces (and also catabolic or breaking down impulses into the metabolism. This leads to thr development of poisons in the plant. The flower is the region of the plant which is closest to the animal-insect kingdoms, and this relationship gives the insects some of the characteristics of the plants and visa versa.

> "If one looks at the butterfly, indeed at any insect from the stage of the egg to when it is fluttering away it is the plant raised up into the air, fashioned into the air by the cosmos if one looks at the plant, it is the butterfly fettered to what is below."

> Rudolf Steiner: Man as Synthesising the Creative Word part lecture iv 26/10/1923.

In some insect/plant pairings this relationship is so well developed that the plants resemble the insect in terms of their flower formation and are specially adapted so that only one species of insect can pollinate them e.g. in types of epiphyte orchids and euglossa bees.

Insects (e.g. moths, bees, butterflies etc.) approaching flowers seem to lose some of their animality and take on some of the characteristics of the flowering world. Conversely, sometimes the plant takes on some of the characteristics of the insects – there is a mirroring taking place. Steiner suggested that the plants and insects were once in the same kingdom, which then diverged. One part became plant, which united itself with the earth element, going towards its origin in the light. The other part, the butterfly still retains the non-terrestrial cosmic structure and has become flower and animal. Thus the insect and plant are like two poles – positive and negative which together become one whole. The flower the mirror image of the negative of the insect, turns to the insect form but also joins with the behaviour and movement of the insects which visit it. There are adaptations on both sides e.g. butterflies with a long proboscis are adapted to flowers with long corolla tubes.

Elementals are primary forces of nature, they make up what is known as the etheric realm of the natural world, and thus sustain all life processes across all other kingdoms. There are four basic types of elementals, earth, water, air, fire.

The Undines (water elementals) are connected with the flowering part of plants and help blossoming whilst the Salamanders (fire spirits) are connected to the fruit of the plant. The Salamanders are intimately connected with insects like butterflies and bees, which are involved in pollination and thereby also fruiting. They actively seek out the auras of these insects to help aid flight, gain more air or astral energy. Salamanders link with the cosmic forces which allow such flying insects like butterflies and bees to remain in a dreamlike metamorphic state of transformation, aligning with the cosmos and the spiritual world more than the earth and the material world.

Of the flying insects there are different degrees of their linking with the higher energies. Those butterflies which pupate on stems and tree trunks are at the highest level of airy detachment from the material world. Other insects eg. beetles and bugs are still very connected to the earth and earth-bound plants and fungi.

Keynote themes of insect remedies

There are keynotes common to insect remedies – which are dependant upon their life patterns, some of these will appear to a greater of lesser extent in the study of the remedies.

Transformation

A lot of insects transform from larvae, or grubs or caterpillars into their adult forms. Hence transformation both in terms of physical appearance and in terms of spiritual development can be seen in some remedy pictures.

Ability to fly

Like the birds, flight is key to a lot of the insect remedies – but here it is often to escape the dirt and mud of the earth, which is where a lot of the insects live. So escapism and dreaming of a better life can be seen. A lot of flightless insects still retain their wings so this 'dream' can still be present. Flight, of course, represents an escape from the earth to either link with the spiritual realm or to link with the cosmic. Certain 'star remedies' (i.e. homoeopathic remedies made from stars themselves such as through telescopic capture of starlight) link to certain insects just as Sol and Apis complement each other.

Materialism

What people look like, what they do, what they own and the whole issue of materialism can often be seen in insect remedies. The insects often have key features of their behaviour linked to their appearance and are trying to escape the lowly jobs of clearing up the mud and detritus. Sometimes appearance is important in insect remedies.

Sexual activity

Many insects are sexually prolific and are also involved in reproduction of plants through their activities. Hence sexuality, as well as infertility issues are often seen in insect remedies.

Hard work / industriousness

Insects tend to be busy and 'work hard' to do whatever it is that they do, so often the remedies show a strong work ethic.

Working together/sociability

Some insects work together in groups e.g. some bees, ants etc and their remedies reflect this with people often giving to the group rather than to themselves. This leads to extremes of caring for others at the expense of oneself (and this is a classic cancer miasmatic pattern). At the other extreme, there are the solitary insects, where the picture is very different, with 'preciousness' and selfishness as the theme. Most insect remedies do however have a 'social' aspect to them in that they care what others think of them.

Vulnerability

Often insect remedies show some sort of vulnerability, insects form a low place in food chains with many predators wanting to eat them. This can lead to 'victim' status – and also restlessness, worrying, irritability, anxiety and even depression.

Fighting and posturing

Insects' behaviour often shows posturing and fighting as a counterbalance to the 'vulnerable' picture. People may show off or even become aggressive and bullying.

Petty, irritating thought patterns / selfish thought patterns

This comes from the esoteric literature – selfish thoughts coming from the lower ego. Petty, irritating thought patterns do show up in some of the remedy pictures. This is because of an ancient belief that insects are actually the physical manifestation of old discarded thought forms.

Energy imbalances

Some insect pictures have issues to do with energy, sometimes related to parasites draining energy out of them, or they drain energy from other people. This can lead to chronic fatigue or a scattered mental energy expended in dreaming. There isn't enough energy to act on one's dreams into materialisation or to physically do what needs to be done.

Cleaning up the planet / ecological concerns

A lot of insects are involved in recycling the energy and physical matter of nature. Through rotting, decaying processes they are effectively cleaning up the planet. This can be seen in some insect remedy pictures with fastidiousness and cleanliness – even going as far as obsessive compulsive behaviour. In others it shows as a huge concern for ecological issues such as global warming, deforestation, pollution, energy crisis etc.

Heat

Insects are attracted to heat and light. Consequently there is 'heat' in the remedy picture, either 'better for heat' or burning symptoms. Often insects emit noxious chemicals that cause irritation or swelling with heat, another clue to the remedy picture of heat. Sometimes insects are so attracted to flames they fly towards, but without what we would consider rational thought Thus 'bypassing rationality' can come out in some of the remedies. However, other remedies have cold features, particularly those involved in clearing up debris in some way.

Duality

Peter Frazer cites duality as part of the insect remedy picture. Seeing things in black or white at one or other extreme rather than as part of the continuum between the two polarities.

METHODS OF PROVING

The remedies were proved in several different ways:

> i) Research into the life cycles and patterns and physiology of the insect itself. A study of related myths and legends where appropriate.
>
> ii) Classical provings – using provers who are not taking other remedies. A regimented protocol of dosage is instituted, keeping a diary log of symptoms.
>
> iii) Shamanic attunement – through taking the remedy in a shamanic ceremony and linking with the spirit of the remedy.
>
> iv) Meditative provings – taking the remedy and meditating on its properties. This included individual and group meditations.

These give an initial remedy picture which was supplemented by using the remedies with clients in 'clinical practice' and seeing which bits of the remedy picture were key to the remedy. This added supplementary information.

THE REMEDIES

1. *Papillon Ulysses*

PAPILLON ULYSSES

Blue Butterfly - Ulysses
Country: Iquitos, Peru
Order: Lepidoptera
Sub Order: Butterflies
Superfamily: Papiliondidea
Family: Papilionidae

BACKGROUND

The 'blue butterfly in my Ayahuasca vision (which took place in the Peruvian rainforest) appeared twice as a dead insect – once by the hut I was staying in, the second time by a tambo a mile into the jungle where I had gone to meditate. It was a large butterfly with very bright blue wings surrounded by black. The colour and size was like a Morpho butterfly but without the white dots often found on the black Morpho butterfly wings. The shape of the butterfly was different from the Morpho butterflies which generally have rounded edges to their bottom wings. The blue butterfly in the remedy had slightly elongated protuberances from the bottom wings and looked like pictures of the Mountain Blue Butterfly. However Papillon Ulysses Ulysses is an Australasian butterfly. Nonetheless, Papillon species are also found in Peru. So I am convinced that the identification of the butterfly is Papillon Ulysses Ulysses rather than a Morpho butterfly. Morphos tend to rest with their wings folded together so the iridescent blue on the wings isn't seen at rest – the blue butterfly in the remedy had come to rest near me in the jungle on several occasions and had rested with its wings outstretched showing the blue. Both Papillon Ulysses Ulysses and Blue Morpho butterflies respond to anything with bright blue colouring.

I still feel it significant that I came to this classification as my journey with the insect remedies began links to the legend of Ulysses, a human hero. This group of remedy provings ended with a large Hercules beetle. Hercules, the archetypal hero whose labours are often used as a metaphor for a human being developing his spirituality and connection with his soul.

CHARACTERISTICS OF BUTTERFLY REMEDIES

Butterfly remedies have the key characteristics of transformation, lightness of being and vulnerability. Often there is an urgency to get things done.

Like other insects, the butterflies have transformed from caterpillars living on the earth and dealing with difficult issues and challenges of surviving in the world. After transformation of some aspect of their life, people needing butterfly remedies recognise they have made a shift in their lives and often feel 'lighter' for that. However, just as butterflies are vulnerable to predators, often these people feel vulnerable in their new position.

Butterflies have a limited life and so a keynote of the remedies is a sense of urgency of needing to get things done quickly, there is a sense of not having enough time. As a relatively long period of their life is spent inside a pupa, there is often a need to pause in life and have the time to just 'be', and to come to terms with oneself. During this time, all the contents of the cells of the caterpillar are completely broken down and reformed into the adult butterfly, a complete metamorphosis and transformation of all aspects of being. This brings in the key aspect to the remedies – they are often indicated for people who are at a crisis stage of spiritual unfoldment. It is at a pivotal moment when impulses from the soul and higher self are starting to come through, causing the lower ego of the personality to be challenged. Often there are lots of small crises in life where the personality wants material things, yet the soul impulse shows up the things that are needed rather than wanted. As the person becomes aware of the conflict between personality and soul, they start to be able to detach and look at their lives, rather like an outsider looking in. This is an early step towards the goal of merging soul and personality.

THE ULYSSES LEGEND

Ulysses is the Greek hero Odysseus, whose history has been compiled by Homer. Ulysses was a hero, but a very human hero, a reluctant hero who never asked to do the heroic deeds but found himself put into situations where he had to rise to the challenge. His motivation was his return to home and family. He was a hero of the Trojan Wars, but was only there because of a promise to a friend. After the war, episodes included various women who kept him away from returning sooner to his wife. He didn't want to be unfaithful, he wanted to go home, but he was tempted e.g. he had himself tied to his ship's mast so he could hear the sirens whilst being able to respond to their lures.

As a remedy, Ulysses Ulysses is a very human person with strong family ties and values. It is someone who is effective in the material world but is starting out on a spiritual path. There is a sense of there being 'more to life than this', of searching for home and starting to transcend duality. It is also useful for people who are isolated in their spiritual search. It can aid them to find other people in their soul group and connect with them.

KEY FEATURES

- Transitions – client going through or has been through a transition or transformation in life
- Lightening – client feels lighter either physically or mentally, feels as though they have let go of something
- Sense of family values
- Allergies and Asthma
- Useful as a carrier remedy – moving the energy of this remedy or another remedy in combination to a specific spot quickly
- Rejuvenation
- Healing the etheric body
- Brain damage
- Attention Deficit Disorder – ADD.
- The need to just 'be' and to re-evaluate ones place in the world

PHYSICAL EFFECTS

- Nose throat, airways, lungs – sneezing, shortness of breath, clearing the passages, asthma (particularly when brought on by allergies)
- Tinnitus – right-sided ear problems, hearing noises
- Cleaning the blood vessels – arteriosclerosis, blood disorders with deposits
- Premature baldness
- Nervous system – nerves firing off and connecting abnormally
- Shoulder and neck problems – right sided
- Speech or throat problems
- Nervous system remedy – for developing brain tissue where destroyed especially in grand mal epilepsy, cerebral palsy and brain damaged children, also after accidents. Helps nerve muscle and mind body co-ordination in children with ADD.

MENTAL/EMOTIONAL EFFECTS

- Lifts you out of the mundane, allowing creative ideas to flow
- Feeling of freedom – having let go of something
- Vulnerability in 'new' phases of life
- Lifts the spirit
- The need to take time out for oneself and just be

- Joyousness, laughter, lightness of being – positive mental plane
- Changes in the perception of thought, by transmuting negative to positive thinking
- Positivity – thoughts and deeds attracting positive outcome and positive situations
- Helps people with 'life changes'
- Relieves stress and anxiety – promotes resilience
- Overcoming duality – brings out 'shadow' things into consciousness for healing
- Things not said – saying what you feel from the heart
- Clairaudience and telepathy
- Wanting to get things done urgently
- Strong sense of family
- Feeling of there must be more to life than this, a sense of isolation from people with similar views
- Men who are family oriented but tempted by other women

EFFECTS ON CHAKRAS / SUBTLE BODIES

- Throat chakra – allowing the throat to open and sing, improving and lightening communication in difficult situations
- Third eye – fluttering
- Crown chakra – clearing
- Brainstem – enables you to feel 'dark' holding patterns there and in the rest of the body – shows up subtle blocks bringing them into consciousness
- Etheric body – heals holes in etheric and strengthens the whole etheric
- Fifth dimension – the remedy links into the fifth dimension and brings in structure

SPIRITUAL ASPECTS

- People who are having life crises because they are starting to feel impulses from their souls which conflict with what their personalities think they want
- People who are starting to develop spiritually
- Remedy links to the innocence of the inner child before any harm came to it

- Adolescence – feeling of freedom, dancing, a letting it all hang out
- Mischievousness, coyote trickster, joking and laughing
- Brings joy
- Aids travelling along ley lines and other energetic grids
- Trust
- Helps thought movement in the brain, to help co-ordinate different brain centres to receive cosmic impulses and channellings
- Aids channelling
- Helps the physical body receive and express a mental state in a beautiful way, as in ballet dancers and body workers.

2. Megasoma Elephas

MEGASOMA ELEPHAS

Elephant Beetle (in rhinoceros beetle family)
Country: Iquitos, Peru
Order: Coleoptera (beetles)
Family: Scarabaeoidae
Sub family: Dynastina

BACKGROUND

This is a large elephant beetle in the rhinoceros beetle family. It was a large black male specimen 12cm long, 5cm wide.

The beetle is heavily armoured with a strong shiny black covering (exoskeleton). It is horned, with which the the males fight one another over females. The lateral horns are short and diverging with a short central horn. There are hairs on the electra. Megasoma Elephas has large wings and is able to fly.

The larvae develop within palm trees and decomposing logs in the Amazonian rainforest. Adults feed at ground level on rotting logs, carrion and faeces.

Amazonian Indian tribes use the horns of the beetle as sacred artefacts to enhance prowess and aid in battle. In modern times, the beetles are used and pinned out for tourists in frames.

KEY FEATURES

- Arthritic hands
- Stress – over-production of adrenaline
- Feeling frozen, unable to move, powerless
- Suppressed anger
- Hot-headed angry males, or females with that type of man in their lives
- Domestic violence

PHYSICAL EFFECTS

- Crushing tension in neck and jaw
- Arthritic hands
- Tending for hand to go into fists
- Liver problems
- Tension in body, fight or flight – overproduction of adrenaline, as a result of stress
- Hot-headed

MENTAL / EMOTIONAL EFFECTS

- Stress
- Hot-headed men – get angry easily
- Women who have above type of men in their lives
- Red mist
- Feeling of being frozen, stuck to the floor and unable to move
- Terror

- Vulnerability
- Cruelty
- Power issues and power imbalances in relationships
- This is a key remedy indicated where there is domestic violence
- Ego-centric people – thinking only of themselves
- Single-mindedness, determination, strong intent
- Armour, protection
- Male sexuality – males either have a strong sex drive and a tendency to not honour the women they are with or they think about sex a lot and may have problems like premature ejaculation or inability to ejaculate

EFFECTS ON CHAKRAS / SUBTLE BODIES

- Solar Plexus disturbances – link to power issues
- Sacral Chakra – for males

SPIRITUAL ASPECTS

- Atlantean karma – particularly pertaining to power issues
- Strong link to the stars – cosmic energies down load quickly

OTHER

- Related remedy - is similar to the 'anger' of lycopodium

(The beetle is sometimes known as dynastes satanus – Satan's Beetle.)

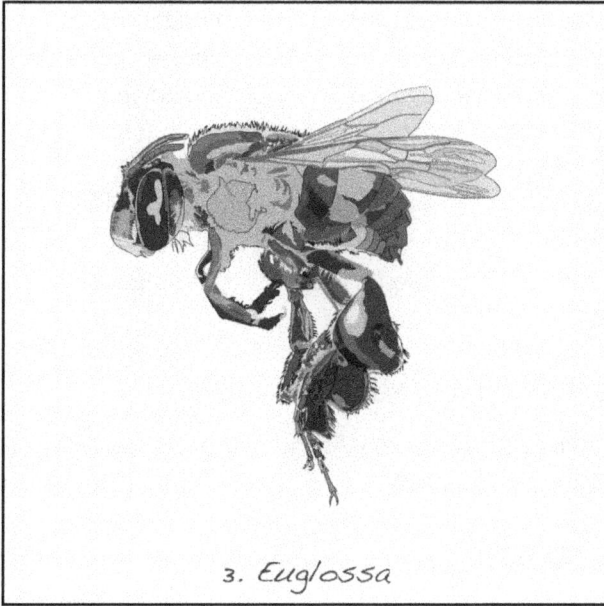

3. *Euglossa*

EUGLOSSA

Emerald Orchid Bee
Country: Iquitos, Peru
Order: Hymenoptera (bees, wasps, ants)
Super: Family Apoidea (bees)
Family: Apidae
Sub family: Euglossinae
Species: Euglossa

BACKGROUND

The bee is a green metallic bee commonly known as the emerald orchid bee. It has an extremely long tongue to get inside the tubular structures of orchid flowers. The females have expanded areas on their hind legs. They are attracted to one particular species of white orchid. The females scrape chemicals off the petals of the orchid with their front feet and transfer this to their hind legs whilst flying in the air. The chemicals are then used in the choice of mate by the males. The males put on group displays to attract females. The males pollinate the orchids.

Orchid bees are sting-less. They are solitary bees who do not swarm or live in colonies – they only come together in the mating season. They make their homes by burrowing into rotting wood and tree trunks in the Amazonian rainforest.

Orchid bees have co-evolved with individual species of orchids and have ingenious adaptation for attracting and removing relatively large pollen packets.

Orchid plants pollinated by euglossa bees are aerial orchids – epiphytes, they are not grounded, but rely on living on the host plant. They are not parasitic i.e. the bee is receiving the solar forces from the mineral earth which gives it the lesson of how to be in unity conscious when working from a dependant stance. The roots of epiphytes are aerial and spongy in nature, they suck in moisture from the air.

> "Humidity is more important to them than warmth. They are found in greatest profusion at those altitudes where the forest is almost continually dripping with moisture. The greatest numbers are undoubtedly to be found in the forests of South East Asia and the northern part of South America."
> Warburg (Die Pflanzenwelt)

In the orchidacae, the epiphytes can grow with their flowers horizontally or even below the level or the roots so although there is a stalk, it isn't strong and upright and is less under the influence of the Sun's vitalising vertical forces than other plants. It has been said by Rudolf Steiner that orchids are all 'flower' (i.e. the over-focus on flower development at the expense of the root and leaf formation). These flowers are highly specialised so that only one species of insect can bring about fertilization for a particular orchid species. Often the flowers have highly specialised mimicry of their animal partner so they resemble their appearance e.g. butterfly, bee or hummingbird. So the orchids have a powerful almost animal astral force exerting its influence in the flowers and the epiphytes are at a higher level of leaving the earth than terrestrial orchid species growing in the earth.

Some epiphyte orchids are sweet smelling and very beautiful, whilst others have monstrous faces and a repellent smell. It is fair to say that 'appearance' is important in orchid remedies and this is carried over to the insect partner, in this case the Emerald Orchid Bee. Orchids are generally not used for medicinal or food purposes.

> Gerbert Grohmann (The Plant, vol 2) states that "orchids renounce
> the mineral element and their etheric plant nature is over-powered
> and absorbed by excessive astralising earth forces."

The Great Orchis means testicles, so links orchids and the animal partners to reproduction systems in associated remedies.

This remedyis not associated with heat but rather with coldness. The lack of heat can be seen in peri-menopausal and menopausal women who are cold and don't exhibit hot flushes, often they feel precious and special and looks are very important to them.

As in all the bees, emerald orchid bee brings in solar forces, but in a different way from other bees.

There is also a picture of people not being grounded and feeling spaced out, sometimes not being very effective in the world. Often people are solitary and not very good at communication and so they can be lonely or isolated.

KEYNOTES

- Menopausal or peri-menopausal women who do not have hot flushes – may be 'cold', often they are precious and feel special (like a princess). How they look, and keeping young and thin is very important to them.
- Ungrounded people
- Sense of isolation and loneliness
- Clears kidneys and ovaries

- Headaches
- Cysts in reproductive system
- Testicular problems
- Issues with taste

PHYSICAL EFFECTS

- Clears kidneys and ovaries
- Ovarian, genital and testicular cysts and inflammation of reproductive system
- Menopause – helps women who are menopausal but have cold invasion and don't get hot flushes to burn out old patterns
- Can be useful in cancer and precancerous states when the warmth of the ego cannot penetrate the cold stagnant parts. The remedy reorganises the 'warmth' in the organism
- Childbirth – helps mother child bonding and the development of the sucking reflex of the baby
- Aching breasts / mastitis
- Kidney weakness
- Low back pain
- Headaches – migraine especially when person is in an uncomfortable situation in life. Headaches are frontal (often near third eye) and sharp and throbbing
- Taste changes, foul taste in the mouth, people unable to taste things
- Meningitis
- Encephalitis
- Hay fever
- Sneezing on waking – or wakes up sneezing during sleep – allergies are often brought about by specific flowers
- Left eye
- Ear wax
- Testicular problems
- Rashes – come out in welts and sting

MENTAL/EMOTIONAL EFFECTS

- Women who feel special or precious
- Over-concern with how they look or staying young
- Solitary people who can be gregarious in crowds but usually distance themselves from people
- Comfort eaters
- Middle-aged women who have gone through relationship break-up but have not regained a sense of their own identity
- Needy women
- Immense confusion about sexual identity – can help homosexuals 'come out'
- Vague, diffused fragmented boundaries with partners in relationships
- Men who hide their feeling behind an outer show of bravado, often they brag a lot and put other people down
- Men who fear failure, need to win at all costs, often they will stick heavily to the game-plan
- Ameliorates male competitive behaviour eg. football hooliganism, yobbishness, boardroom fights, power struggles
- Feeling of being 'weighed down' often by life
- Suppressed anger –when it comes out people quiver with anger and have a tendency to lash out and bully (useful to follow with homoeopathic Berlin Wall)
- Obsessive, compulsive states – particularly to do with cleanliness
- Slows people down who are manic
- Combats feelings of aloneness
- Scatteredness from disassociation and ungroundedness (not the scatteredness of conventional apis which is about flitting from one thing to the other)
- Fear of what other people think, getting old, losing their looks

EFFECTS ON SUBTLE BODIES/CHAKRAS

- Linked to the 5th Dimension – structure
- Balances heart and solar Plexus chakras
- Helps ameliorate solar plexus upsets due to power struggles
- Sacral chakra – helps bring in cosmic impulses to replace old links to Earth and Moon in menopausal women

- Base chakra – issues with grounding – strengthens meridians in the thighs
- Kidney chi deficiency with cold invasion

SPIRITUAL ASPECTS

- Atlantean and Lemurian karma
- Helps support people whose cords to the heart chakra get stretched during astral travel in their sleep and who feel ungrounded on waking. Helps strengthen the cord and bring them back into their bodies if taken on wakening.
- Links to parallel lives
- Visions of Sphinx, fish, buffalo
- Increases creativity, particularly in menopausal and post-menopausal women – dance, art, music etc
- Spiritual awareness of lower ego reality
- Invites souls to reach for and merge with the personality to overcome the sense of separation and link with unity consciousness

OTHER

- Sycotic miasm – can be secretive
- Useful for people who have multifaceted problems, particularly those to do with fear
- People who need to get way to find solitude or who have hidden themselves from the world
- Symptoms ameliorated by going for long, slow completive walks in nature

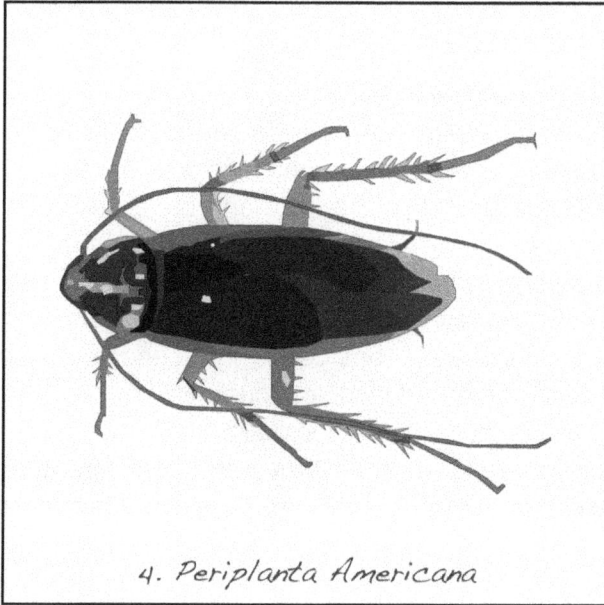

4. Periplanta Americana

PERIPLANTA AMERICANA

Giant American Cockroach (Female)
Country: Iquitos, Peru
Order: Dictyoptera (Cockroaches and mantids)
Sub Order: Blattodea
Family :Blatteria
Genus: Biaberus

BACKGROUND

Cockroaches are pests, that finds homes where there is warmth and an abundance of food. They are omnivorous and will eat dead animal material including dead cockroaches but will not kill each other for food. They are nocturnal and effective at hiding during the day so they can build up to large numbers before being noticed. They have two pairs of net wings and can fly but prefer not to, as they are more adapted to running rather than flight.

This gives the remedy picture the feature of wanting to escape from earthly, mundane things, but often not succeeding in that and so there is a sense of losing out, of frustration. The remedy picture is of a nocturnal person, often staying up late, past midnight and having difficulty getting up in the morning. They are often very motivated in the workplace but sometimes even though they are often successful they feel a sense of unworthiness and guilt and are over-concerned with what other people think of them.

Cockroaches are resilient, they can survive radiation, and not eat or drink for several days. Their bodies have a covering of wax and a long, shield-like protective layer, which helps conserve moisture even in very dry conditions. Their harm does not lie in the food they eat, but in the fact that they spoil it by contaminating it with a noxious odour. (Provers found that their feet smelled or that they became sensitive to odours!)

Female cockroaches lay eggs in a purse-like container, which protrude from her abdomen. These purses are deposited once they have formed. The nymphs initially are worm-like, but quickly shed their skins and emerge as tiny cockroaches. They have five 'shedding' stages before reaching full size, a process which takes several months. The provers found a sense of dissatisfaction with work, especially early on in their career – they may have had several jobs before settling on what they wanted to do.

In South America these cockroaches are sometimes eaten so there is a sense of fear and a need to get away on ones' own in the remedy picture.

KEY FEATURES

- Asthma
- Lungs
- Thirst
- Successful at work

- Over concern with what others think
- Wanting to escape the mundane, but not being able to
- Guilt
- Smelly feet or sensitivity to odours
- Nocturnal – often stay up after midnight and can't get up in morning

PHYSICAL EFFECTS

- Wet asthma – worse in damp weather
- Chronic cough – phlegm and mucous
- Lungs opened up to chi especially when congested with mucous and debris
- Helps move energy from lungs to the feet
- Clears sinuses
- Frontal headache, sensitive to light head feels like it is bursting
- Migraine
- Helps sodium processes in the cells
- Skin – helps clear skin and complexion, often there is a tingling sensation as the nadis open up
- Thirsty
- Spinal degeneration
- Dementia
- Clears bowels of putrid food
- Anal problems – tissues and piles
- Constipation with stool going back in
- Cramp in top of feet
- Smelly feet
- Oversensitive to smell

MENTAL/ EMOTIONAL EFFECTS

- Successful at work but is over concerned with what other people think and there is a sense of guilt, of not being good enough. Often there have been several other jobs before doing the present job.
- Frustration – wanting to escape the mundane but unable to do so
- Bereavement, loss – especially where there is clutter and festering of the emotional life
- Good for clearing unfinished business

- Selfish people, getting what they want
- Letting other people fester
- Enjoy moving to a new area
- Like change
- Survivors
- The remedy can work deeply into murky dark emotional areas to clean out debris
- Likes the dark
- Likes being alone
- Need to get away
- As if imprisoned
- As if the light is fading or there isn't enough light
- As if surrounded
- As if constricted
- Tendency to get things out of perspective
- Big things are trivialised; little things assume huge importance
- Anally retentive of ideas

EFFECTS ON CHAKRAS / SUBTLE BODIES

- Opens third eye
- Move chi down from lungs
- Lungs open up to new chi
- Clearing of spinal column via waves of Metatron particularly magnetic wave
- Well-developed mental body

SPIRITUAL ASPECTS

- Helps clear clutter in the rear reptilian brain and at the 4th, 5th and 6th cerebrospinal ventricles
- People who are trapped by memories, possessions and relationships – they can't find any basis of reality apart from through material things
- Helps souls let go of other souls – even at the level of past lives enabling new soul connections rather than repeating patterns

OTHER

- Compare with Blatta Orientalis, Ahriman, works well with Sri Maha Bodhi - a remedy from the leaf of the Sri Lankan tree which was propagated from the original Bodhi Tree (Fiscus Religiosa) under which Buddha gained enlightenment.

s. Euschistus Heros

EUSCHISTUS HEROS

Brown Stink Bug
Country: Iquitos, Peru
Order: Hemiptera (Land Bugs)
Sub Order: Heteroptera
Family: Pentatomidae (Shield Bugs)

BACKGROUND

The Brown Stink Bug is a type of 'shield' bug. It has a broad-shouldered look, with a triangular shield (scuttellum) between the wings. The edge of the insect has lighter /darker markings in sequence.

The female fiercely guards her nymphs, she lays clutches of about 60 eggs about five times a year, and it takes about five weeks after hatching for the nymphs to become adults.

The adults are strong fliers and readily fly over long distances. They are major pests of crops including soya beans, grains, fruits, nuts etc. They feed by inserting needle-like mouth parts and releasing toxic substances which stop growth and may cause damage to structures e.g. tillering. It also can cause an effect like a cat's face where surrounding tissue grows around damaged tissue. It can inhibit root growth, grain production and make the plant susceptible to pathogens.

The Stink Bug is given its name because when threatened it emits a noxious odour including. Octyn-l-yl acetate (2E)2, decane, uridecane, tridecane

KEY FEATURES

- Women who are very family orientated, protective of children
- Chronic fatigue syndrome
- Helps relaxation in the muscles
- Post operative remedy – clears anaesthetic and analgesia
- People constantly moving on
- Lung problems
- Putrefaction, purulent states
- Irritation

PHYSICAL EFFECTS

- Itchy scalp
- Left eye irritation and styes
- Shoulder blade stiffness – the remedy aids relaxation of muscular tension, particularly in the neck, shoulder and intercostal muscles
- Opens sinuses and clears out etheric mucous from brain if head is feeling fuzzy
- Itchy rashes
- Lungs – helps clear alveoli, relaxes tension, eases spasms, opens chest, ribs and breast bone, aids deep breathing
- Tired – yawns a lot
- Chronic fatigue syndrome, particularly when there has been a happy, protracted childhood and where the person has not been equipped to face the world.

- Loss of sensation to the hands
- People who have had strokes or accidents, where the mind is active but the body is not
- Wounds which are festering, smelly and putrid
- Clears tumours (gonad, adrenal, kidneys) which have associated pus or infection
- Worm infestation causing adrenal deficiency
- Helps metabolism which is immature e.g. putrefaction or fermentation from improper breakdown or catabolism of proteins
- Trichomonas and other sexually transmitted diseases, especially if with odious discharge
- Odious foetid feet

MENTAL / EMOTIONAL EFFECTS

- Good for drug users and alcoholics
- Vulnerability
- Detachment from people and situations, but in a cheerful way
- People who can't cope with life and build up walls to protect themselves or become addictive
- Independent, keeps other people at bay
- Feels spaced out
- As if hearing other peoples' thoughts
- Problems with boundaries to other people
- Women who are very family orientated and protective of their children
- Feels as if tongue is too long for mouth
- Aids creative problem solving
- Irritation with others
- Helps people with slow mental processes
- Sense of 'someone looking after me' of being shielded and protected
- The remedy can build resilience; bring a sense of joy, laughter and acceptance of the situation.

EFFECTS ON CHAKRAS / SUBTLE BODIES

- Heals gaps in the aura, bringing people back to themselves and gives them boundaries
- Links to astral and 4th dimension, helps emotions to flow
- Clears out debris in all subtle bodies
- Helps energetic pathways within the body

SPIRITUAL ASPECTS

- Spiritual boundaries – brings clearer sense of self/not self
- Aids contact with nature spirits and body elementals
- Can aid communal awareness
- Helps bring in new spiritual blueprints through the lungs, then moves them to the brain and down throughout the body
- Helps aid creativity
- Removes old memories and wounds, helps to erase the past

OTHER

- Works well with nux vomica
- Men with broad shoulders, T-shaped physique
- Related remedies: pyrogen, calendula, carcinosin
- Sycotic miasm – with forgetfulness
- Syphilitic miasm

6. Lethoserus Annulipes

LETHOCERUS ANNULIPES

Giant Water Bug
Country: Iquitos, Peru
Order: Hemiptera (True bugs)
Sub Order: Heteroptera
Family: Belostomatidae

BACKGROUND

Lethocerus annulipes is a large water bug 145 mm long. It is brown, oval and flattened with a face resembling a 'grey' alien.

It lives in stagnant ponds, and can assist the recycling of debris from them. They also inhabit slow moving streams, as well as being found near water. They cannot breath under water, but have two short tubes at the tip of the abdomen to exchange air with a large air bubble under the wings. Air enters the body through holes in the abdomen called 'spiracles' and they swim using modified hind legs resembling oars. They feed on insects, small fish, tadpoles, frogs, water-creatures and snails – mainly live specimens but they can eat dead as well. The forelegs have curved hook-like claws to grasp and hold their prey. They then bite the prey using a needle-like beak and inject their prey with a poison containing tissue destroying enzymes. They then suck the tissue fluid of their prey.

They are known as 'toe biters' because they occasionally attack swimmers, biting humans if they are threatened or are hungry. When grasped they eject fluid from their anus.

Lethocerus annulipes are attracted to lights and become disorientated. They navigate using fixed points of light, such as stars therefore other light sources interfere with their natural navigational systems. They are sometimes known as 'electric light bugs' because they are attracted to lights.

Lethocerus is considered a 'delicacy' in Asia, but is also eaten in South America.

Reproduction: Males flap the water creating waves to attract the attention of females. Females lay eggs on vegetation above the water, but sometimes lay them on the backs of males. The eggs hatch in 1-2 weeks, with total development time to young adulthood taking about six weeks. Adults sometimes hibernate for periods in mud.

This gives a remedy picture of aggression and vulnerability and in a way links to the new homoeopathic remedy of 'Berlin Wall' and its power struggles. They also show disorientation and confusion about their spiritual development and can get involved with glamour and illusion, eg. using magical means to manipulate others or to get things for themselves, or feeling they are great healers etc. This can link into the area of 'curses' and black magic and this is an appropriate remedy to help with those issues.

Their adaptations for breathing can help clients with breathing difficulties particularly those with lungs filled with fluid or a collapsed lung, but is not a general lung remedy.

KEY FEATURES

- Aggression
- Vulnerability
- Victimisation
- Power struggles
- Bowel and colon problems
- Confusion about spirituality
- Curses and black magic
- Lungs – fluid filled or collapsed
- Childhood tantrums
- Light sensitivity
- Abuse of sexual power
- Fear of memory of alien abduction

PHYSICAL EFFECTS

- Feels 'thick-headed' – frontal region and forehead
- Eye problems – weak eyesight, sharp pain at the back of eyeballs
- Light sensitivity – in eyes or on skin
- Soreness or ulceration in right corner of mouth
- Lungs – hard to breathe, as though full of water, fluid filled or collapsed lungs, helped by hot steamy atmosphere
- Oedema
- Lower back pain
- Joint and spinal problems – strengthens back
- Arthritis and rheumatism – particularly affected by wet conditions and humidity
- Atrophy and wasting away of upper arms
- Disinterest in sex
- Abuse of sexuality with control issues
- Clears anaesthesia
- Bowel and colon problems – with runny, watery diarrhoea, queasiness and a sense of wanting to bring something up, to regurgitate

- Bulimia
- In vitro fertilisation – particularly children born through IVF with weak kidney energy due to genetic damage
- Cloning

MENTAL / EMOTIONAL EFFECTS

- Aggression
- Power struggles
- Victimisation
- Vulnerability
- Apprehension
- Feeling 'I don't know where I am'
- Going through the motions, but feeling "I might as well be dead'
- Fear of or memory of alien abductions
- Naughty children
- Childhood tantrums
- Sado-masochistic tendencies – leather and bondage
- War, destruction, devastation, terror
- An obsessive need to chase after one's objectives and goals even when they are not on the right path
- Compulsive need not to give up
- Cold emotionally
- Sexual abuse – victim and abuser
- Frigidity after trauma
- Rigid, analysing, limiting and controlling people
- One track minds
- People who need to get away and find solitude and just be with themselves in the face of hectic situations
- Happy to go along with things to help keep the peace
- Emotional blocks
- Sense as if drowning
- Feeling spaced out
- Woozy feeling – as if head is being flooded from top down
- Exhaustion
- Can give clarity and sense of direction
- Can give someone grit, backbone or determination
- Mentally opens up clarity of though, awareness of the choices

ahead, clears the clutter getting in the way
- Helps mental tiredness

EFFECTS ON SUBTLE BODIES / CHAKRAS

- Opens blocks in heart chakra
- Calms Solar Plexus, helps to balance it and raise its energy to the heart centre
- Suppressed sacral centre
- Kidney chi weakness
- Strongly etheric remedy – when the etheric has got out of control and there is chi stagnation and a lack of flow at any level.
- Third eye chakra 'swimming' – difficulty in 'perceiving' things
- People who are trying to be true to something and can't
- Brings rhythm and flow of spirit through the chakras

SPIRITUAL ASPECTS

- Helps anchor destiny and life purpose
- Duality
- Atlantean karma – particularly power struggles
- Lemurian karma – particularly sexual deviance
- Can provide a platform to access other dimensions
- Clears karma
- Disorientation and confusion about spirituality
- Helps aid 'soul retrieval'
- Black magic
- Curses
- Someone who is in a spiritual trap and held in glamour and illusion eg. someone who thinks they are 'the best' healer, false gurus, people using spiritual means for selfish ends
- A separation between brain and body with blockages in the brain stem

OTHER

- Sycotic miasm
- Tubercular miasm – feeling more in the present, bringing them into the 'now' less projecting into illusory states

- Links to 'greys' dark brotherhood, conspiracies
- Alien abduction
- Planetary travel
- Zetareticuli extraterrestrial karma – including of genetic diseases, bringing an end to the race – couldn't carry on
- compare with Berlin Wall for aggression and power struggles
- compare with Tarantula for mental aspects

7. Comahen

COMAHEN

Tree ant
Country: Iquitos, Peru
Order: Hymenoptera (ants, bees, wasps)
Sub–order: Apocrita
Family: Formicidae
Genus: Cephalotesatratus

BACKGROUND

This is a homeopathic remedy made from a Peruvian herbal remedy called Comahen. It is made from a cone taken out of a tree ant's nest, and included eggs, grubs, ants, and parts of the nest. It is boiled in water for two hours, strained and added to honey and lemon. It was bottled and the bottles were buried underground for 9 days and nights. This allowed the mixture to ferment. The herbal remedy is used for bronchitis, TB and other lung conditions, particularly those where there is damage to the alveoli. The belief of the Shipibo Indians is that when the ants repair the damage to their nest the person's lung condition will be improved.

As a homeopathic remedy, the honey adds the warmth of the sun to the remedy picture and the lemon brings some of their qualities of the citrus remedy. However most of the remedy picture is from the ants and the ants' nest.

Lemon brings in the circulation and can improve poor circulation and lymph drainage and any oedema. It also includes the throat to the general remedy picture. Note how with honey it is often given to relieve sore throat symptoms. The honey itself brings in the quality of 'liquid sunlight' and directly links to solar forces. It is also a powerfully acting natural antiseptic and wound cleanser and as a homeopathic remedy is indicated for wounds, cuts, stings, burns, sores and ulcers and for skin complaints such as eczema and psoriasis.

Although these are seen in some of the provings and clients cases, the major part of the remedy picture is that from the ant's nest and the ants, eggs and grubs themselves. This obviously has links to the Formica rufa remedy, but is somewhat more complex.

The tree ants are a very social community. They are black and about double the size of most British ants. The ants nest is placed on trees in the rainforest above head height. The ants themselves move down the tree and forage on the rainforest floor, bringing back different materials for the nest and food such as larvae and caterpillars. They move outwards from the tree communicating by the means of pheromone chemicals and their tracks can last up to five days for the ants to find their way home, even in the rainy season.

The fact that these ants live in trees rather than on the ground means that they are more in tune with air and therefore the 'lungs' than the focus on joints and bones found in our usual Formica rufa. Hence in this remedy lung complaints are of greater importance than its use in arthritis, rheumatism, gout and polyps found in Formica rufa. The astral forces are stronger in their sphere of action within the lungs – whereas in Formica rufa, the interplay between the astral and etheric forces takes place in the kidneys. In Comahen the classical migratory arthritis are still features but of secondary importance to the lung features. The modalities can be similar with hot swollen joints with may become suppressed and develop over time into cold deformed joints, migratory pains changing position and intensity, sudden onset pains worse for cold and damp, better for motion, warmth and pressure. There are also premalignant moles, lesions and polyps, and other rheumatic disorders such as gout.

Ants recycle old unwanted material and ensure that this is in a state to be reused by the ecosystem, especially utilising formic acid metabolism. The human formic acid process ensures that the DNA-RNA constituents of proteins and purines do not break down into toxic ammonia and excess uric acid, which can otherwise lead to gout or contribute to formation of polyps and other nonmalignant growths. Thus a key feature of any ant remedy is to recycle old tissue e.g. malformed arthritic joints, polyps, persistence of scar tissue. Formic acid processes prevent toxic build up in the tissues. With tree ants, this helps remove old tissue from lung damage. These processes are essentially astral catabolic or 'breakdown' forces which then allow the etheric anabolic or 'building up' forces to regenerate new tissue.

Ants in particular represent the irritation of petty low-level human thoughts and how these cluster together and create more negativity in the world - as represented by the irritations of biting ants when they are disturbed.

The worker ants are infertile females. The males impregnating the queen do not otherwise have an important role in the colony. This gives a picture of workaholic females wanting to get on with their career, whose work is more important than a relationship. Or they try to have children late in life but then develop infertility problems. They often carry heavy loads at work. Males on the other hand, will not be work focused.

The tree ants are wingless but use their flattened head, middle and hind legs and flanges to help them to perform acrobatics to turn up to 180°

in the air and glide back to the tree trunk if they are blown off it. Such manouvres meet with 85% success (5% would be the statistical norm). As they live on tree trunks, they are more highly evolved than ants living in the earth. Their ability to glide and turn in the air can reflect in the remedy profile with patients who have already started down their path of spiritual development.

KEYNOTES

- Lung problems, particularly those with physical damage, emphysema, TB, pneumonia. Also works with bronchitis, pleurisy asthma etc
- Inability to let go
- Grief
- Negative petty thought processes
- Workaholic females
- Women who have infertility problems after leaving it late to have children
- Purification of mind and body

PHYSICAL EFFECTS

- Lacrimation, runny eyes
- Cough with phlegm – coughs at night, wakes up coughing
- Worse cold and damp motion, over-exertion – better for warmth, rest and pressure
- Dry sore throat
- Wandering, migratory, radiating symptoms
- Mucus, catarrh, phlegm
- Lung problems – bronchitis, emphysema, pleurisy, chronic obstructive airways disease, TB, pneumonia
- Polyps – nasal and elsewhere, e.g. colon, rectum, cervix
- Rheumatism and arthritis in joints with wandering symptoms – migratory arthritis starts with hot swollen joints then over time becomes cold and deformed.
- Pre-malignant lesions, cysts, polyps, dysplastic moles
- Ovarian cysts
- Infertile older females

- Males with difficulty ejaculating
- Remedy assists the immune system phagocytes in scavenging old pus and tissue cells, recycling obsolete tissues especially within lungs, soft tissues and muscles

MENTAL / EMOTIONAL EFFECTS

- Inability to let go
- Petty negative thought forms
- Rigidity of mind – recycling thoughts that can't unwind
- Females focused on work – driven looking for success
- Males not focused on work
- Feeling of 'being burdened' - 'carrying a heavy load'
- Grief
- Living in the past – regrets for past actions
- Sociability, craves company
- Likes memorabilia – may collect things
- Nostalgia – wants to return to 'old ways'
- Craves sugar
- Remedy helps bring clarity of mind – helps focus on individual thoughts and see how they affect each other and people and things around them
- Females can have tremendous highs on new relationship – get very carried away and ecstatic, but then lose interest quickly
- Males have 'one night stands' and are afraid to commit, may have phobic/paranoid feelings of being killed or dying after ejaculation so may have difficulty ejaculating

EFFECTS ON SUBTLE BODIES / CHAKRAS

- Weak mental body – lots of petty negative thought forms but not very focused
- Astral body stimulated in lungs to destroy unwanted energies and allow new blueprints into the new tissue
- Etheric body stimulated to grow new tissue – this is a strongly etheric remedy
- Stuck energy in throat and heart chakras
- Wind cold-damp invasion leading to solidification of mucus into pus

SPIRITUAL ASPECTS

- The remedy activates the inherent intelligence of the body to detoxify and breakdown obsolete tissues in the physical and to detoxify the mental body of negative stuck thought forms
- It brings in flow to the 4th dimension and allows for re-building and bringing in structure in the fifth dimension – with new cosmic 'Adam Kadmon' blueprints
- The remedy brings a deep sense of peace and a balanced response in the mind and emotions to outside triggers

OTHER

- A strong sycotic miasm remedy
- See also Thuja, Formic acid, Formica rufa

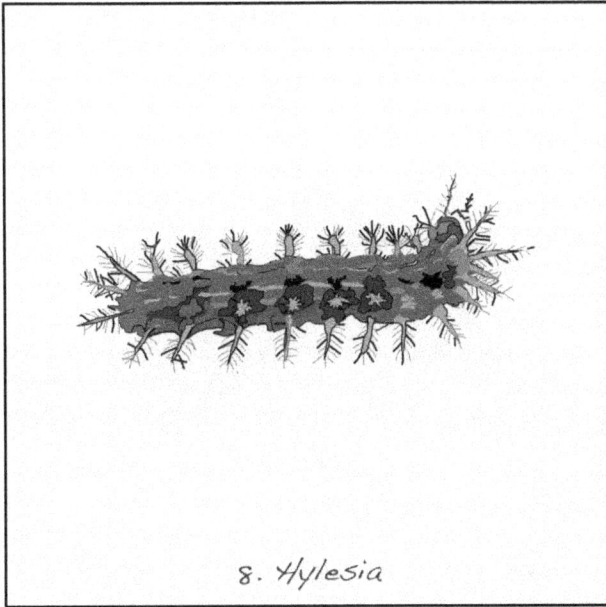

8. Hylesia

HYLESIA

Caterpillar of Hylesia moth commonly known as BAYUCCA in Peru
Country: Iquitos,Peru
Order: Lepidoptera (butterflies, moths)
Sub Order: Glossata
Super Family: Bombycoidea
Family: Satunidae
Species: Hylesia

BACKGROUND

The caterpillar was brought to me with a client suffering from a raised red rash on his face which was swelling up. He had brushed past the caterpillar which was on the leaf of a tree at head height. He was very agitated because when people in the village had touched poisonous caterpillars before they had respiratory difficulties and blood disorders where there was internal bleeding, bruising and disruption of the blood clotting mechanism. The swelling on his face was hot, painful and throbbing. I removed two caterpillar spines and gave him Apis immediately, then made the caterpillar into a remedy and gave it to him at 4x and 30c about an hour later. This stabilised his symptoms, the 4x working on the physical and the 30c working on his 'mentals' as it decreased his agitation. The swelling, which started on his cheek, was so bad that within an hour it had swollen so his right eye could not open. The pain was sharp and unbearable on touch and was building up and getting worse. It was very red and mottled. He felt very hot although was not sweating. We made a compress of cold water with the 4x remedy in it and put it over the swelling, which relieved some of the pain.

He then went to hospital for an antihistamine injection as there was no anti-venom for Hylesia. The trip took about 2 hours. Hylesia caterpillar stings have been known to kill people.

Saturnidae is a group of large moths with large, hairy bodies known as Emperor moths, Moon moths or Royal moths. They have well-developed eye spots and transparent patches on both fore and hind wings. The adult moths either have no tongues or small vestiges of tongues so are unable to feed and die 5 days after coming out of the pupa. All adult Saturnidae moths have markings like eyes on their wings. They have a very sensitive sense of smell and can find mates over large distances utilizing large antennae.

The caterpillars tend to follow each other using pheromones secreted by their anal glands, so they are sometimes know as 'procession' caterpillars and are often found in groups.

The caterpillars are hairy with spines. They moult about five times before pupating. The spines are hollow and are attached to a single basal poison cell in the caterpillar. These fracture on contact with the skin to inject the poison under pressure, sometimes the spines can remain in the skin causing localised reactions. Some caterpillars release setae into the environment, these hairs can remain poisonous for several years and when inhaled can cause rhinitis or respiratory disease.

The poison contains histamine releasing substances, 5 hydroxytryptophan and other components, some which cause haemolytic, proteolytic or hyaluronidase activity. In some patients immunoglobin E (IgE) antibodies are produced on contact resulting in severe allergic reactions and hypersensitivity.

KEY FEATURES

- Red swelling with inflammation and oedema, constant pain unbearable on touch
- Allergic reactions – severe
- Urticaria
- Anti-coagulant
- Numbness
- Stings
- Anorexia

PHYSICAL EFFECTS

- Red swelling, inflammation and oedema, (may be blistering) constant pain, unbearable to touch
- Toxic dermatitis, erythroderma
- Severe allergic reaction – sometimes on the skin but can also cause throat and lips to swell up and block airways leading to unconsciousness
- Numbness – often moving down the body but can occur anywhere – feeling of local anaesthesia.
- Urticaria
- Rhinitis – with red eyes or swelling around eyes
- Acne rosacea patterns
- Red faced people with round faces which look like they are ballooning
- Bleeding problems – difficulty in clotting, anticoagulant states
- Anorexia
- Stomach problems with discomfort as if 'hungry' with feeling of emptiness
- Want food but get problems if they eat so irritation or diarrhoea
- Bowel and colon problems particularly useful if people have had colostomies and have a colostomy bag

- Kidney weakness, adrenal deficiency
- Boils, abscesses
- Hot at night – needs to sleep naked
- Absence of sweating
- Stings

MENTAL / EMOTIONAL EFFECTS

- Disorientation
- Fear
- People who hold their emotion in check, but suddenly have a 'sting in the tail'
- Extreme irritability – nothing is right – but this is often hidden under a cool exterior
- "Yes but" people
- Anxiety
- Want to be free but are also frightened of freedom
- People with boundary issues both with other people – they can encroach or be encroached upon, but also people who pick up on emotions like anger or fear from outside themselves
- Pain in the physical connects with astral, emotional pain from old hurts and wounds – this can bring old cellular memory up to the surface
- Numbness of emotions, or people who distance themselves from their emotions

EFFECTS ON SUBTLE BODIES / CHAKRAS

- Deals with misalignment of the subtle bodies
- Pain in the astral body
- Solar plexus
- Fear affects the kidneys and weakens kidney chi

SPIRITUAL ASPECTS

- Extreme irritation leading to 'imperil' which affect energy flow through the energy meridians and can have effects on the physical and mental bodies
- Brings 'out of body' experiences and a dream or trance-like state

- Gives a jump-start to spiritual process
- Brings astral energy into the physical very quickly
- Can be used to lead in 'the fire of spirit' to burn away obsolete energies and memories, especially in 'cold' people. This can kick start a healing reaction in interior cold diseases e.g. precancerous polyps and tumours

OTHER

- Compare with Apis

9. Batus Barbicornis

BATUS BARBICORNIS

Long Horned Beetle
Country: Iquitos, Peru
Order: Coleoptera (beetles)
Super Family: Cerambysidae (long horn Beetles)
Family: Cerambycinae
Tribe: Trachyderini

BACKGROUND

This long horned beetle was black with six red rectangles on its body, red antennae the length of its body with four 'furry' black bits on each antennae. The antennae are olfactory organs which track pheromones for mating purposes. The adult long horned beetles feed on flowers, but the larvae are serious pests because they bore under the bark of trees and destroy them, the larvae can also eat leaves.

The key features of the long horned beetle is the length of the antennae, and as a homeopathic remedy, much like antelope horn, it is about 'receiving' messages from the spirit world of divine guidance in a practical way to give practical solutions to down to earth problems. The guidance is from the cosmos, and the higher mind at the buddhic and atmic spiritual dimensional levels.

The adults' attraction to flowers brings a sweetness into the remedy, either of relative contentment with their life or as a craving for sweet things, cakes, biscuits, sweets, sugar, chocolate. Often they have had more difficult childhoods where they have had to struggle to survive. Sometimes there will be a history of childhood anger and tantrums with destruction of toys etc, possibly by Attention Deficit and Hyperactivity Disorder, ADHD.

KEY FEATURES

- Migraine – throbbing, pulsating headache in frontal region near temples, with light sensitivity and nausea (especially in school children and students)
- Aftermath of strokes with one-sided bending of the body (left side came out in the provings)
- Rigidified thinking
- ADHD, destructive tempers in children
- Poor circulation due to hardening arteries
- Worry

PHYSICAL EFFECTS

- Over sensitivity in senses, tinnitus, vertigo, light sensitivity aggravated by movement
- Hot, burning ears
- Throat constriction
- Clears airways in sinusitis, colds, cough, flu etc and balances inhalation and exhalation
- Circulation disorders

- Hardening and stiffness – client 'feels as if hardening' either of body parts, bones or hardening of the arteries
- Kidney disease
- Shoulders and spine out of alignment, may be anti-clockwise movement, turning and hoisting 'feels like a screwing movement', like moving round a wormhole in space
- After a stroke or accident, bending to one side, especially left
- Spleen weakness

MENTAL / EMOTIONAL EFFECTS

- Childhood anger and destruction, ADHD, temper tantrums
- Adult contentment often after difficult childhood
- Anger, irritation
- Worry combined with rigidified thinking where illusory realities have been created from excess worrying and fretting – at an extreme this could lead to an obsessive / compulsive pattern
- Rigid thought patterns
- Glamour and illusion
- People who clown around putting a smile over a sad face with hidden grief, sadness, separation
- Male – warrior complex having to fight for things and be strong
- Childhood nightmares – monster under the bed
- Fear of miscarriage
- Panic attacks
- Childhood / student headaches related to study
- People who won't take advice \ People feeling dirty and want to be clean

EFFECTS ON CHAKRAS / SUBTLE BODIES

- Crown chakra opening, aligning with higher chakras – reaching to the cosmos, receiving divine guidance from cosmos and from higher self at buddhic and atmic levels. This brings practical solutions to every day problems.
- Balances in and out breath and the astral body coming in through the lungs which helps the letting go of old memories and the in coming of new impulses from the cosmos

- Third eye partially blocked with narrow perspectives
- Grounds astral through kidneys
- Spleen yang deficiency and dampness in spleen leading to worry and headaches

SPIRITUAL ASPECTS

- Enhances trance states
- Enhances psychic abilities
- Clairaudience
- Links to angels
- Helps to see ahead – as if there is a beam of light in front of you bringing things into focus and seeing aspects that may have been hidden
- Links to 2nd star in Perseus – Mirfak may bounce from there to other star systems. Mirfak influences North American Indian shamanic traditions particularly cleansing rituals e.g. sweat lodges. It helps clear rigidified thinking, often headstrong people who won't take advice – which is a keynote of this remedy for young people either through sweating or cleansing with water or by bringing in Shiva, solar flare forces to clear the way for new energies, i.e. cleansing by fire.

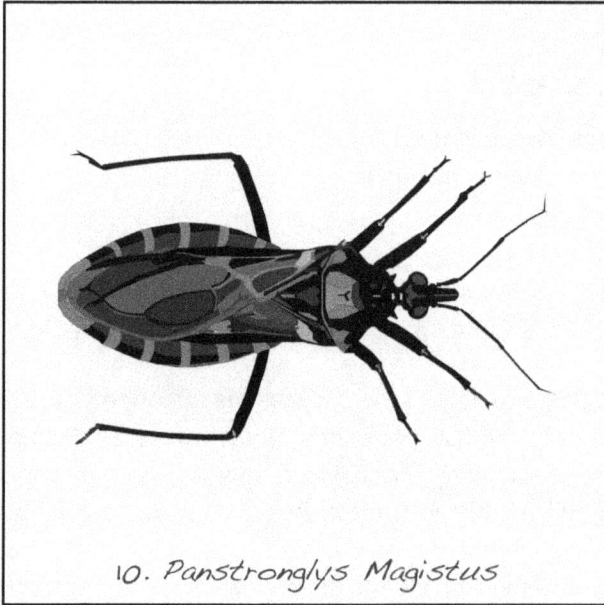

10. *Panstronglys Magistus*

PANSTRONGLYS MAGISTUS

Reduvid Kissing Bug (carrier of Chagas Disease)
Country: Iquitos, Peru
Order: Hemiptera (True Bugs)
Suborder: Heteroptera
Family: Reduviidae

BACKGROUND

The kissing bug, Panstronglys magistus lives in the walls of houses in
South America, particularly where there is poverty and deprivation. It is
called the 'kissing bug' because it comes out at night and bites people
typically at the corner of the eye or lip. Kissing bugs carry the parasite
'Trypanosomas cruzi' which causes the sleeping sickness 'Chagas
disease'. The parasites are single haemoflagellates which live in the

blood or lymph of the host, they are deposited as faeces on the skin as the reduvid bugs feeds and are rubbed into the skin if the irritation due to the bite is scratched or are rubbed into the membranes of the eye, mouth and nose.

Mild forms of Chagas Disease include swelling and reddening of the site of the bite followed by the swelling of the eye and swollen lymph nodes often with lymph oedema. The liver and spleen also become enlarged. Often the initial phase of Chagas disease is followed by remission for fifteen or more years, after which is may relapse. Symptoms may then manifest of cardiomyopathy and digestive abnormalities, as swallowing difficulties and can lead to malnutrition or infestation of the colon causing severe abdominal pain and constipation.

KEY FEATURES

- Sores – nose, eyes, corner of mouth
- Cracked skin on face, hands, feet or heels
- Obsession – after making up and believing that the object of the obsession loves them, with a hatred for other people around that person
- Inability to let go
- Counteracts mental confusion or side-effects caused by opiate analgesics
- Metabolic toxicity of heart muscle-cardiomyopathy
- Sleepiness, drowsiness, narcolepsy
- Insomnia
- Chagas disease
- Bites
- Localised redness around wounds
- Constipation – hard dry stools following gut / intestinal problems
- Dementia

PHYSICAL EFFECTS

- Works to balance sleep patterns, useful for sleepiness, drowsiness, narcolepsy and insomnia
- Sores near eyes, nose, mouth with localised redness

- Dry cracking sores or dry cracked skin on face, hands, feet or heels
- Bites and puncture wounds
- Lymph problems – enlarged lymph nodes, lymphatic congestion, lymphoedema
- Metabolic poisoning of the heart muscle, cardiomyopathy
- Arrhythmia, tachycardia, congested heart failure
- Pancreatitis
- Gut and intestinal problems with undigested food leading to chronic constipation with hard dry stools
- Dryness
- Chagas disease
- Bitter saliva
- Swallowing difficulties
- Dementia
- Parasitic infection
- Worms
- Vitamin B deficiency – cracking at corner of mouth
- Herpes simplex
- Dry eyes leading to soreness – problems with tear ducts
- Dark red blood – heavy blood, high iron content

MENTAL / EMOTIONAL EFFECTS

- Obsession – focussed on one person, makes up stories (and believes them) that their affection is returned, can be hatred of people around the object of the obsession
- Inability to let go – hangs on to old patterns, relationships, jobs
- Hatred or self hate
- Dementia
- Confusion of mind – the remedy is indicated when drugs ie. opiate painkillers are affecting the mind bringing lack of focus or confusion
- Too tired to bother
- Night people
- Difficulties in establishing boundaries with other people

EFFECTS ON CHAKRA / SUBTLE BODIES

- Allows energy flow between eyes and liver

- Blocks in heart chakra
- Third eye chakra
- Tears in etheric body near stomach and intestines
- Karmic cords from the intestines
- Energy imbalances with other people – either feeding off the other person's energy or the other person is feeding off their energy (energy vampirism)
- Person ungrounded and out of their body, with a lot of energy in astral body – creating 'unreal' scenarios and obsessions
- Balances heart yin and small intestine yang pair of linked organ energies

SPIRITUAL ASPECTS

- The remedy helps ground people and bring them back into their bodies
- It balances the two spheres of the brain releasing and opening up the different subtle bodies and promoting energy flow through the physical body
- It helps to define 'self 'and 'not self' and to establish boundaries to all the subtle bodies
- The heart cannot connect to the spiritual self so the soul feels trapped within the earth and the subconscious realm. Ultimately the remedy invites you to keep your power and not overly nourish or support others in a dependent way. It does this by balancing the chi between the heart yin and small intestine yang chi. These people have excessive karmic cords arising from the intestines and are feeding souls stuck in hell or purgatory or with criminal tendencies. They have had past lives where there was much trauma in the gut and intestines, so there are fears in the etheric body. This results in the metabolism being stuck with dense underworld energies and in an inability to let go and a thickening of the blood, also the heart chakra becomes heavy, dense and toxic particularly in the heart muscle. This toxicity results in self-hate or projected hate in the emotional body, which makes the saliva bitter.

OTHER

- Sensation as if the rib cage opens up with relief of tension
- Related remedies:
- Alumina (dementia, dryness, constipation)
- Tuberculinum, Opium, Cannabis indica (mental symptoms)
- Ledum (puncture wounds)
- Holly flower essence (hatred)

II. *Erotylus*

EROTYLUS

Pleasing Fungus Beetle
Country: Iquitos, Peru
Order: Coleoptera (Beetles)
Sub order: Polyphaga
Super family: Cucujoidea
Family: Erotilidae (flower, bark and ladybird beetles)

BACKGROUND

Erotylus, the Pleasing Fungus Beetle is the size of a ladybird, black with thin, serrated orange markings on its back, the orange colour changing shade from nearly yellow to nearly red going along the body. Erotylus is a beetle which reproduces profusely, giving rise to its name. It lives on the rainforest floor and feeds on various fungi in the rotten wood. Its' colouration is a warning signal for predators, once attacked it exudes a noxious liquid from its anus and organs on its legs.

KEY FEATURES

- This is a major reproductive remedy useful to balance sexuality, for infertility and general sexual problems
- It is also a key remedy for fungal infections especially under the toe nails, thrush in the mouth and candida
- It is helpful in cases of past sexual abuse with guilt and fear
- In vitro fertilisation, IVF
- Herpes
- Candida
- Cancer of reproductive system

PHYSICAL EFFECTS

- Nose bleed, with sinusitis
- Ears popping – especially right ear
- Can help eye problems because it brings 'moistness', it is useful if there is a decline in eyesight with a subsequent increase in 'spiritual sight'.
- The remedy brings moisture and is useful for any 'dry' conditions.
- Headache – dull ache, back of head
- Herpes
- Skin problems including psoriasis and herpes
- Never well since abuse
- Allergies to do with excrement, e.g. house dust mite
- Fungal infections – candida, thrush
- This is a major remedy for toe nail fungus which softens it, but needs to be followed by another remedy to draw the infection out. It can be used both orally and topically.
- Reproductive problems, including:
- Infertility – helps prepare the womb for pregnancy / in males helps ease anxiety and brings relaxation to reproductive organ to sallow them to function effectively.
- Menopause – with dryness of tissues, premature menopause
- Vaginal discharge – including where the microbes are not isolated or unknown even after testing
- Brings vitality to the reproductive system
- IVF and other infertility treatments

- Polycystic ovaries
- Activates oil-making processes in metabolism of ovaries (thus promotes oestrogen and progesterone) and adrenals (androgens)
- Ovarian deficiency syndrome
- Premature aging
- Problems with testes
- Helps increase libido
- Pelvic inflammatory disease, particularly if there has been a history of promiscuity or multiple sexual partners
- HIV-AIDS
- Chronic fatigue in men with a background of multiple partners, and excessive seminal loss
- Kidney deficiency states
- Liver and bile duct disease – especially if linked to eye problems
- Irritable bowel disease, IBS and other gut problems including malabsorbtion and constipation with dry stools, light in colour.
- Crohn's disease
- Cancer – particularly cancer of reproductive system – with tissue growing out of control e.g. fungal – mushroom or condylomata like overgrowths
- Ichthyosis
- Movement is a feature of the remedy – either it is held rigidly in check and stilled or is frenzied. Symptoms are better for movement in the first case, worse for movement in the latter

MENTAL / EMOTIONAL EFFECTS

- Misuse of power – sexual abuse
- Saying 'no' meaning 'yes'
- Fearing intimacy masked by promiscuity
- The remedy brings personal self empowerment within sexual and love based relationships and enables intimacy with the self, leading to intimacy with a trusted partner
- Surfacing of memories of past sexual partners particularly if there have been issues of abandonment, unrequited love, fear of intimacy, guilt, obsession
- Person 'lost' in their sexuality – men obsessed with sex without thought for feelings of partner or even who the partner is and women

who are obsessed with the man and will do anything to please him
– even if there is something telling them to hold back or they don't
feel right about it
- Women who have gone off sex after childbirth or during a long
monogamous relationship
- Men who boast about sexual conquests
- Helps remove guilt and fear especially around sexuality
- Gender insecurity
- Issues with intimacy and fear of intimacy, especially when hidden by
a joking façade, sarcasm and cruel jokes, which are hurtful to others
- Manipulative, controlling people
- Victimised people who hide this by an outward show of bravado
and false strength, often where others say something like that (e.g.
abuse) could never happen to her
- People who desire relationship at any cost
- The remedy enables people to open and balance their hearts
- Needy women who can be clingy
- Inner male and inner female sides, to come to terms with their own
sexuality and intimacy. This new balance enables them to honour
themselves and take that into a sexual relationship, where they
are not overly needy on their partner and can allow openness and
intimacy into the relationship.

EFFECTS ON SUBTLE BODIES / CHAKRAS

- Sacral chakra – this is re-balanced with energy flows from the adrenals
and the kidney system to vitalise or revitalise the reproductive organs. The
block in the sacral are overcome, and this allows energy flow between the
sacral and throat chakra so that people can express themselves.
- It helps release karmic cords from the sacral centre, and allows creative
flows from here
- The remedy balances the etheric body and prevent excessive overgrowth
of tissue. It repairs holes in the etheric and astral bodies due to trauma and
abuse and works well with Sequoia (redwood tree) to repair boundaries
and ground the energy.
- Base chakra – grounds the energy and strengthens the system
- Increases kidney chi and help it flow to revitalise and balance the
reproductive system

- Heart chakra – can help remove blocks which come from past abuse or past sexual partners
- Brainstem – helps remove old implants from ancestral patterning
- Movement can either be help rigidly in check or become frenzied – the remedy helps to bring the 'movement' impulses which originate from the astral body back into balance
- Helps repair holes in the aura
- Blocks in crown and third eye chakra related to pineal and pituitary which result in excess of congested energy in the gonads

SPIRITUAL ASPECTS

- Brings to the surface old sexual cords which need to be recapitulated, and old karmic cords from past lives. It also brings up unresolved issues of unrequited love, obsession, abandonment and rejection, particularly when these have been held in the cellular memory of the reproductive system.
- Excessive activation of forces of fertility in a forced way e.g. infertility treatments, IVF where ovarian stimulating drugs are given
- Karmic clearing – particularly Atlantean karma related to abuse of power and Lemurian karma related to sexual experimentation
- Can help link to souls wanting to incarnate
- Activation of blueprints in eggs and sperm, bringing in higher spiritual forces and anchoring them in the physical plane through natural conception
- Brings in unconditional rather than conditional love in relationships
- Brings in the 'light' to 'dark' aspects of past and present relationships enables people to see aspects to issues which have previously been hidden
- Awareness of past lives and past relationships which have a bearing on relationships in the present
- Brings in joy
- Helps people to be present and 'in the now' in sexual relationships
- It can help remove blocks and to come to a sense of oneness with acceptance, surrender and inner calm
- Grounds people into the Earth Grid
- Balances Hermes / Aphrodite energies in the brain and brings inner male and female energies into balance

- Links to Milky Way Galaxy, which is a gateway to letting in star energies particularly at the menopause when old links to the Earth and Moon have been burned away by the Fire of Spirit, allowing reconnection and deepening connection with cosmic energies

OTHER
- Syphilitic, AIDS and sycotic miasms
- Related remedies:
- Thuja (fungal infections)
- Adventurine (opens heart in sexual relationships)
- Agnus castus (chronic fatigue related to seminal loss)
- Staphysagria (enables ownership of your own sexual energy)
- Medorrhinum (fear / guilt)
- Works well with:
- Redwood (fungal infections)
- Sri ma Bodhi / Bodhi Tree (fungal infections)
- Scorpion (female victims of sexual abuse)
- Berlin Wall (power issues – both victim and victimised)

12. *Nagicicada Cassini*

MAGICICADA CASSINI

Cicada
Country: Iquitos, Peru
Order :Hemiptera (True Bugs)
Suborder: Homoptera (cicadas and plant hoppers)
Family: cicadidae.

BACKGROUND

This cicada was a large cicada, body size 3½ inches long with two pairs of transparent wings which completely covered the body. Cicadas are strong flyers. The forewings are hard and membranous providing a strong structure for flight. They feed on the sap of trees and plants with wide beaks arising from the back of the head.

The constant whistle-like 'singing' of the cicada is produced by two membranes called tymbals in cavities on either side of the abdomen, the action of tiny muscles produces the resonant sound, the abdominal cavity amplifies this. The male cicadas sit in trees or bushes and practice the sound, which is a mating call, the sound is almost constant and can be deafening if a lot of males are present. Song increases with heat and they like the heat. Each cicada species has a distinctive song and pitch. In addition to the mating call some have a quieter 'courtship' song once they have attracted a female. There is also a distinctive distress call, which is disjointed and emetic, which occurs when a predator has seized a cicada. The females have an ovipositor, which cuts slits into trees and shrubs, and this can be quite destructive. The name cicada means 'buzzer'. Cicadas have prominent eyes set wide on the sides of the head with short antennae set between the eyes.

Cicadas face the same direction during mating. The females lay several hundred eggs in trees or shrubs under the bark. After hatching, the nymphs drop to the ground where they burrow and feed on the roots. The nymphs have enlarged front legs for digging. They go through seven moulting stages and spend a long life phase as a nymph e.g. 13 or 17 years in the magistus genus, 7 years in the Tibicer genus. The time as a nymph is always out of step with the 3 or 5-year cycle of its predators e.g. killer wasps, as a way of preserving species numbers. The adults emerge from the last moult, which takes place on a plant, which the nymph has climbed. They have a relatively short time as an adult.

Cicadas are known as 'dry flies' because of the extreme dryness of the skin left on moulting. They can cool themselves by sweating, and can raise their body temperature by as much as 22° celsius above the ambient temperature.

Cicadas are mainly eaten by birds or cicada killer wasp, but they also commonly die from massospora cicadina, a fungal disease. Some cicadas are eaten by fish if they fall into ponds or streams. Cicadas have a defence mechanism, the number of cicadas in any area by far exceeds the number of predators. The predators are thus satiated leaving the remaining cicadas to breed.

Cicadas are a 'delicacy' in many parts of the world e.g. China, Malaysia, Burma, Latin America, the Congo, and in ancient Greece.

In Taoism, cicadas are the symbols of hsien, when the soul disengages with the body at death. In Greece they were sacred creatures of Apollo.

Cicadas are often included in jolt-law e.g. in France they represent insouciance (nonchalance or indifference), this is represented in La Fontaine's fable 'The cicada and the ant' where the cicada sings all summer whilst the ant is busy storing food, so she finds herself without food in the winter. In China the phrase 'to shed the golden cicada skin' is used to represent the tactic of using deception to escape danger, particularly using decoys or likenesses to fool enemies. In Chinese mythology the multiple shedding of the cicada shell symbolises the many stages of transformation required of a person before all worldly illusions have been broken and enlightenment is reached.

KEYNOTES

- Nerve paralysis
- Parkinson's disease
- Strengthens & revitalizes muscles – skeletal system
- Liver or speech problems
- Someone stuck in earlier stages of life
- Birth trauma – e.g. umbilical cord round neck
- Congestion & constriction
- Females who are verbally destructive and girls who we are best friends with one minute and fighting the next.
- Worry
- Rave music and drug culture
- All symptoms are characterised by being persistant & constant, better for heat

PHYSICAL EFFECTS

- Birth trauma – e.g. umbilical cord around neck
- Congested headache – like having a disc-like band around and

within head, constant, not letting up, better for heat, aggravated at night
- Hot in the head
- Eyes
- Poor vision
- Cataracts
- Seeing double in vision & floaters
- Hardness, e.g. of skin, joints, blood vessels
- Dryness generally
- Alzheimer's disease
- Cracked dry skin
- Lips, e.g. cracking of
- Heel disease, e.g. cracking, calcaneal spurs
- Eczema & psoriasis
- Scabs or sores and wounds
- Musculoskeletal system disorders, e.g. arthritis and rheumatism

(The remedy strengthens the musculoskeletal system generally. It can be used for overworked and strained muscles or muscular degenerative disorders. It is helpful for people with muscle wasting in their legs, from underlying neural-muscular disease or from long-term immobility and convalescence.

- Bones – it can help strengthen the bones, particularly in the legs, cases of osteoporosis and brittle bone disease
- Spleen – the remedy can be helpful for people with spleen problems, (in Chinese medicine the spleen nourishes the muscles and blood with food related chi and governs muscles tone)
- Liver – imbalances can cause muscle spasms and cramps, especially within the digestive system, also tendon disorders such as contractures and tendonitis
- Although dryness characterises the remedy, in two cases taking the remedy resulted in profuse sweating which was odourless

MENTAL AND EMOTIONAL ASPECTS

(These are often ameliorated by music)
- Male bravado – being one of the boys, making lewd jokes, being

noticed in all gatherings e.g. down the pub, but this hides a gentle side to the character, who can be vulnerable. Sometimes males are very conscious of their appearance

- Females that have destructive tendencies – throw things if angry, but are more destructive verbally, making comments to deflate and undermine other women particularly. It can be useful remedy for girls who are best friends one minute and fighting the next
- Female irritation – verbalises negative thoughts forms, nag people
- Vulnerable, feel under attack, believe as though someone is talking about them, or giving them 'dirty looks'
- Nonchalant, indifferent people
- People who are stuck in earlier stages of their life, in childhood patterns, e.g. don't want to get old and take action to avoid this e.g. make up, clothes, having a younger girlfriend, cosmetic surgery, HRT
- Depression: follows lithium mica or lepidolite well
- Worry
- Useful to help with 'milestones' in life, transaction and change
- Someone who is too busy having fun to secure for their future
- Feel hot, hot-headed, hot as if wrapped up or in a blanket
- Feels as if head is separate form body or head being suspended in space
- Feeling of whirling anticlockwise, like being in a vortex or wormhole
- Feels like a dried up old shell
- Desire for a large family (or from a large family)

EFFECTS ON CHAKRA / SUBTLE BODIES

- Can only communicate to the world through the head, through the crown, third eye (ajna) and throat centres – through sound and intention of thought
- Body is cold, devitalised and shut down, chakras below throat are often not in alignment and shut down, base chakra is not really grounded into the earth
- The sacral and throat centres can work well together after taking the remedy, this is stimulated by music (sound baths, Tibetan bowls, gongs) and produces sweetness and joy throughout all the subtle bodies

- Third eye opening or increase in all round spiritual sight

SPIRITUAL ASPECTS

- This relates to a time in human evolution when the body was still being developed – as briefly recapitulated by the embryo stage or young baby who lives largely in the head
- The multiple shedding of the cicada shell represents the multiple stages of transformation required before overcoming the illusion of the human world and gaining enlightenment
- Useful for musicians to help improve the quality of their music and align it to spiritual principles
- Links to a star with insect-like beings who are at war with the machines

OTHER

- Compare with arsenicum iodatum (congested headache), lithium mica (lifting out of depression)
- There are strong links between this remedy and club culture of the 1990s and early 2000s with techno dance music, strobes, flourescent lights and raves. Typically there is some drug use – usually ecstasy, and larger than life clothes e.g. platform boots, spiky hair or shaved heads
- Some provers saw the remedy being linked to insect-like beings such as those from the star Rigel in the constellation Orion and those taking control of machines like beings e.g. those that we created on Procyon in Canis Minor who later enslaved beings on formalhaut

13. *Enallagma Cyathigerum*

ENALLAGMA CYATHIGERUM

Blue Damselfly
(From Cornwall) – seen May – September
Order – Odonata (Dragonflies)
Sub order – Zygoptera (Damselflies)
Family – Coenagriidae

BACKGROUND

The common blue damselfly is a narrow winged damselfly with one stripe on the side of thorax. They hold their delicate, membranous wings together above their body while at rest. The males are blue; females can be blue, green or brown.

They live near ponds, streams and brackish water. They are particularly skilful fliers and can hover in the same place for a long time and perform aerial acrobatics. They can beat their wings up to a hundred times a second and reach speeds of up to 60 mph. They can reverse direction mid air in the distance of the body length.

The adult damselflies are territorial and can aggressively defend their territory. They eat insects and are predatory. During mating the males clasps the female to his reproductive organs forming a 'mating wheel'. They fly together to a suitable plant in the water and the eggs are deposited on the plant below the water. The nymphs, called naiads have pincers to grasp their food. They are carnivores and feed on insects and small fish. In turn, they are preyed upon by other adults. When they mature they climb up a stem of an aquatic plant and pupate. The adult damselfly emerges early in the morning. This transformation from the ugly slow moving nymph to the beautiful, ethereal adult takes place in the late spring. The adult phase is short lived and last only a few weeks to a few months.

Damselflies have large compound eyes which dominate their heads and give all round vision, almost 360 degrees.

The common blue damselfly has been recorded at a higher altitude than any other damselfly or dragonfly.

There is a strong connection in folk-law between the dragonflies and snakes. In welsh they are called gwas-y-neirdr (the adders servant) and in South America they are said to live near to snakes, and are believed to be their eyes and ears. However, the same is said of the red dragonflies and it is more likely to be them rather than the damselflies. Some Amazon rainforest tribes link the red dragonflies to evil spirits and are very superstitious about them, not wanting to kill them, believing that bad luck would come if a red dragonfly came near the house. There were similar superstitions about snakes with red and black markings. From an esoteric point of view, damselflies are highly evolved insects who have succeeded in their quest to fly and leave the material world behind them. They appear to be a good remedy for someone who is working on their own spiritual development.

KEYNOTES
- Eye problems
- Degenerative diseases with optic symptoms - spinal problems – particularly above waist level
- Rapid movements of the extremities e.g. tics, shakes, tremors, Parkinson's disease
- Babies with problems in skull e.g. after forceps delivery or non-closure or too rapid closure of skull bones
- Difficulty distinguishing left and right, dyslexic or dyspraxic
- Spiritual people, often with refined vibrations

PHYSICAL ASPECTS
- Eye problems, e.g. blurred vision, Horner's syndrome, 'floaters' in the eyes, photosensitivity
- Multiple sclerosis affecting neck and brainstem
- Spinal problems which restrict movement particularly those above the waist
- Neck stiffness
- Helps align and straighten back, helps stand upright, assists physical atlas realignment techniques
- Helps decongest maxillary sinuses
- Babies with skull problems delivery or non-closure / too rapid closure of skull bones
- Headache – frontal and sides or circling the head, can feel like a band round the head. The pain is dull and aching and is better from sitting or standing still and upright
- Ears – helps decongestion after colds and flu
- Mastitis
- Degenerative diseases e.g. optic neuropathy, multiple sclerosis related optical symptoms
- Rapid movements of the extremities, tics, shakes, tremors, Parkinson's disease
- Helps to work with body shape
- Hungry and or thirsty

MENTAL AND EMOTIONAL EFFECTS

- Clean, tidy, organised, proud – could be a good housewife
- Difficulty distinguishing left and right – may be dyslexic or dyspraxic
- Concerned with appearance – well groomed and well turned out
- Jealousy
- Predatory, territorial people who can wait a long time to pick a moment to pounce
- People who stand their ground and justify their position even though they know it is wrong
- Help work through feelings of anger and to resolve old disputes and fights
- Vulnerable, very sensitive emotionally, swayed by what other people think or say or what they think other people have thought or said about them
- Thirsts for knowledge (going along with physical hunger and thirst)
- Clairvoyance and or clairsentience
- Helps old memories to surface
- Feels the need to dance
- People who have two paces to life – full on and stop, they are busy people who pack as much as they can into a day, they switch to doing nothing and just 'being' (this may be as a result of a life crisis or a new life choice)
- Aids bringing you into a quiet mental state ready for meditation
- Helps you to become present and live in the moment
- Aids being able to review the past with detachment
- Brings clarity of thought
- Feels he might die young
- Males can be assertive sexually
- Feeling of being out of body – "I'm beside myself", "my heads not with me today" etc
- May 'remember' past lives

EFFECTS ON SUBTLE BODIES / CHAKRAS

- Opens the crown to allow in cosmic thought forms particularly in males
- Activate the pineal parietal eye

- Activates ability to decipher channelled messages within the third eye chakra
- Opens up head and neck chakras, increases flow of energy and csf flow in the spine (through alta major and medulla oblongata), rebalances energy within brainstem
- Aligns chakras along spine
- Can awake the kundilini energy
- Aids astral flow
- Out of body experiences

SPIRITUAL ASPECTS

- Helps to activate the various energies in the connecting matrices or grids around the head and neck e.g. translator orbs, third eye facilities and clears head and throat chakras to treat degenerative disease, spinal and eye problems. It also helps balance the energy in the brainstem to help neck and spinal problems
- Can bring an awareness of past lives and their relevance to today's life
- Aware to the spiritual history and past biography of the soul is stored in the head
- Brings an awareness of this for himself and others
- Helps clear memory and old ancestral or past life patterns out of head and neck
- Links in with Mary (mother of Jesus)
- Links to angels of light
- The colouration of dragonfly signifies that during the adult fertile stage, the males are more in tune with cosmic energies and thought forms and are concerned with a 'mental' approach where as the females are more linked to the earth and past and to the moon energies
- Damselflies link to snakes and can help to raise the kundalini
- Can help access the fifth dimension and restructure parts of the body

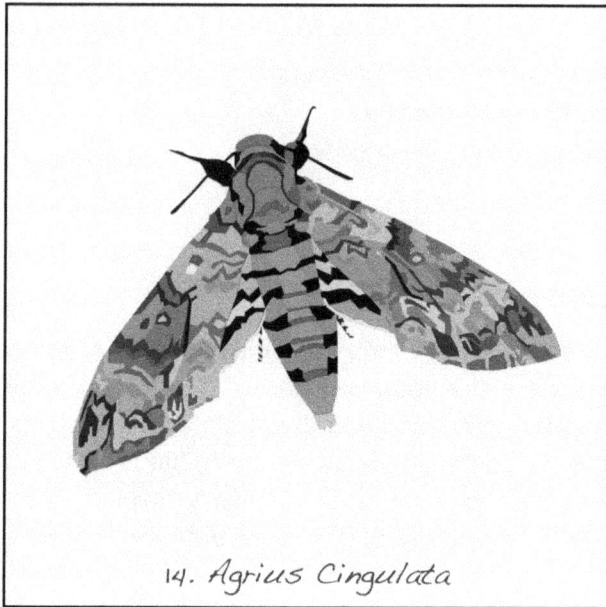

14. Agrius Cingulata

AGRIUS CINGULATA

Pink Spotted Hawk Moth
Country: France
Order: Lepidoptera
Sub order: (butterflies and moths)
Family: Sphingidae, sphinx moths, sometimes known as hawk moths
Genus: Agrius

BACKGROUND

The pink spotted hawk moth is usually a neo-tropical species but has been moving into Europe and been found in Portugal in recent years. This specimen was found in France in 2003. They live in open areas of low land.

They are well camouflaged and are active during the day. Adults are powerful fliers with a wingspan of 3-5 inches, they can reach speeds of up to 30 mph. They can hover which gives them the name 'hawk' moth, they sometimes feed on the wing. Adults migrate in large numbers once a year and gather in the evening before the migration. Their tongues are well developed and they feed on the nectar from deep throated, tubular flowers particularly jimsonweed/thornapple (datura), bindweed (convolvulus), moonflower (calonyction aculeation) and petunia. They have stout bodies which taper to a point, with narrow pointed fore wings like arrowheads at rest. They have large compound eyes,

Their bodies are grey with pink bands and they have vaulted grey and brown forewings with grey hind wings with black bands and pink base. They are well camouflaged against bark.

Females release a pheromone from a gland at the tip of the abdomen, which attracts the males. After mating she lays her eggs on sweet potato (ipomoea batatas) or jimsonweed (datura) or related plants. The eggs are about 1mm in diameter, are translucent with greenish tinge and are almost spherical.

The larvae go through three instars stages in a short time, they can take as short as three weeks to reach maturity. The first instar has a white body with black anal horn; the second and third instars are green or brown with prominent black slanted markings on each side of the body with a black anal horn. The head is green or brown with three dark stripes on either side. The caterpillars, also known as sweet potato hornworm, are large and feed day and night – they are major pests of sweet potato crops. A major problem is parasitation of the larvae by microplitis of the Braconidae family and Metavoria of the Tachinidae family.

The caterpillars pupate in the chambers dug underground and the pupae resemble pitchers and are 15mm wide by 55mm long and reddish brown in colour.

KEY FEATURES
- Sleep problems in childhood
- Constantly hungry

- Detoxification of vaccines
- Premature ageing
- Dizziness
- Sinusitis
- Bulimia
- Taste changes
- Children who are forced to grow up quickly and never had a proper childhood
- People who live life in the fast lane and are always on the go
- Nervous restlessness and sleep problems
- Self criticism, self harm (particularly in children)
- People who have forgotten how to die, lingering in suffering states of end-stage disease
- Anxiety

PHYSICAL EFFECTS

- Lack of sleep in childhood
- Headache with pounding ears, feels pressured
- Sinusitis with severe pain of stuck mucous
- Eye strain
- Pins and needles – nerve endings tingling
- Constantly hungry – babies, toddlers, teenage boys who will eat and eat and eat
- Taste changes, or cannot taste food
- Fatigue – tiredness with yawning
- Bulimia
- Premature ageing
- Dizziness and fainting
- Lack of sleep with nervous restlessness
- Digestive problems – particularly in intestine, better at night, better from being outside and putting your feet on the earth
- Stiffness of or injury to back of neck and or shoulders with resultant tension
- Self harming (cutting)
- Detox vaccines

MENTAL / EMOTIONAL EFFECTS

- People who are active at night and don't need much sleep – (this can tip into insomnia at the extreme and that isn't usually perceived as a problem until later)
- Lack of sleep in childhood
- Assertive females who 'wear the trousers' at home
- Children who are forced to grow up quickly and never had a proper childhood
- 'Streetwise' children
- Old before their years
- People, who camouflage themselves, don't like to be seen and fade into the background
- Emotions are felt – but at a distance, and they are not owned or dealt with
- Feeling of floating
- Feeling of things spiralling out of control
- Anxiety
- People, who have lots of ideas, do lots of seemingly creative or worthwhile things but seem to get nowhere. They hover at the same point and expend a lot of energy going nowhere
- Grounded, sensible, conventional people
- Self criticism – may lead to self harming
- Appreciation of beauty especially feminine gossamer like fragility
- Loves music - music can transform into positive energy and help evaluate symptoms

EFFECTS ON SUBTLE BODIES / CHAKRAS

- Balances solar plexus
- The remedy helps pull energy from the lower chakras to the head chakras
- Helps to ground energy into the earth grid
- Head centres blocked from reviewing new cosmic influences
- Blocks or cords in the heart chakra
- People who have forgotten how to die
- Attachment to material things

SPIRITUAL ASPECTS OF MOTH REMEDIES

Moths are attempting to reach the butterfly state of transformation and 'flying' of the mundane earthly world into the spiritual world, however they have not quite got there. Trapped thought forms are holding them back. This often shows as a kind of 'nervous restlessness' with astral and ego forces stuck in the physical realm within the nervous system, unable to leave the earth. This shows in the behaviour of the moths near lights or flames and can show up as restlessness as the moth throws itself into the flame to bring about the transformation. Also as the way moths hover by flowers to pollinate them (unlike butterflies which usually rest onto the flower during pollination).

It can also show as 'sleep' problems. In sleep, the spiritual and astral bodies travel to the spirit world but are still attached to the physical body by silver thread, the physical and etheric bodies are regenerated whilst the higher bodies return to their original planes, astral 4th, mental 5th, spiritual 6th. Where people are all well connected to the higher planes during the day, they need less sleep at night.

However the characteristic of the moths of not being able to sleep is also there, in this case either in childhood or at the extreme case of the workaholic adult who had existed on little sleep for years and then finds him / her self exhausted and unable to sleep. A lack of sleep indicates an inability to let go of something in the material world or someone who is very materialistic which lets the astral and ego forces get stuck in the nervous system and be unable to leave the body at night. The lack of sleep itself can lead to the physical and etheric bodies becoming depleted, as they are not getting enough sleep to regenerate themselves.

Although most moths are nocturnal, some are active during the day and these timings often tie in with the activity of the client in some way, particularly if he exhibits particular symptoms or behaviour patterns at certain times.

In moths the wings are not joined so they expend more energy to fly than do butterflies, and dark colouration helps them to gather heat onto their bodies to bring in some of that extra energy. This is particularly important in northern climates.

Often in moth remedies, people are grounded into the earth plane and are so connected that they cannot leave it although they want to. So there is often a search for spirituality, a feeling of being lost or disconnected and ultimately this can lead to people living beyond their allotted time span because they are still 'searching' and have forgotten how to die because they are so attached to the earth and material things. Again, this can be seen as nervous restlessness.

Moths often are camouflaged, and in patients there are hidden aspects to their stories or they dress in a away so as not to be seen. This is in a contrast with the butterflies where how they look and 'being seen' are important. Butterflies are much more susceptible to peer pressure than moths.

SPIRITUAL ASPECTS OF PINK SPOTTED HAWK MOTH REMEDY

Workaholic adult, works well at night needing little sleep. Sleeplessness in childhood. Nervous restlessness and sleep problems. In adults sleeplessness related to this remedy are due to them not having let go of something during their childhood or teenage years and connected to the spiritual world. If they have done this successfully, they leave the stage of insomnia and nervous restlessness and become the hardworking types requiring little sleep who are connected to the spiritual world and higher plane during the day. However some people are still stuck in the stage of searching for spirit, with trapped thought forms and attachments to materialism of the earth plane itself, this can lead to nervous restlessness and insomnia in adulthood. Some children and adults know on some level that they need to 'let go' and connect with the spiritual world and self harm by cutting themselves to release old memories through the blood, and become very self critical as result of their inability to let go. This often is an unconscious rather than conscious impulse.

Camouflage is another theme of the remedy – something that has been hidden – has been brought to light in the client's history, or you may find that the client is emotionally or physically trying to not stand out in a crowd. This maybe seen in their clothing style. In the case of a moth, the pink colouration indicates that the 'hidden' aspects are related to heart chakra blockages or attachments.

The well-developed tongue to pollinate 'deep throated' flowers e.g. datura and the hawkmoth brings in the question of the relationship between the moth and flowers. The flowers themselves are firmly rooted in the earth and are not able to leave it, but in attracting the moths which can fly in the air element they can attract some of the energy, which is in the moth's aura. Sometimes taste changes, or lack of taste can be related to the remedy.

Hawkmoths are related to plants of the Convolvulaceae family, the bindweeds grow in a spiral pattern with weed stems which spiral around the plants. So they are more 'mercurial' than related to the sun forces of other plants, mercury orbiting in a 'spiral' pattern around the sun. They also have strong leaves and the flowers come from the joints between the leaves and stems rather than from the end of the stems as in most flowers. These flowers also exhibit a spiral tendency because they are formed progressively in an anticlockwise direction. Goethe regards the winding plant as 'searching externally for which it should give itself but cannot'. Uber Die Spiraltendenz der Vegetation.This means that the flowers are more related to mercury and the moon than they are the sun. (Some bindweed flowers even open at night for pollination). The convolvulaceae and other plants with deep-throated flowers particularly in the solanaceae family, bring in an excess of astral forces to the relationship with the moth during pollination. In some cases, e.g. datura this excess astrality shows itself in the 'poisonous' nature of the plant itself – the poison being used medically in the indigenous cultures. The moth takes on board some of these 'astral' catabolic breaking-down forces during pollination, this can link with the self criticism and self harming children who can be helped by the remedy.

The pink spotted hawk moth is more grounded into the earth plane than many other moths because it pupates in underground chambers in the earth itself, rather than pupating on a plant or tree ie. above the earth. In reacting to the air element this adds to the restlessness characteristic of moths, particularly in the case of this remedy in childhood, because the moth has such a short time in the larval stage. This can bring about an anxiety to take the next step. This often means difficulty in letting go of attachments or material things; this is either taken up successfully giving an effective 'workaholic' successful adult, or instead the adult state is restlessness, having lots of ideas but not being able to realize them, or having sleep problems into adulthood.

The pink spotted hawk moth is a sphinx moth and this brings in a sense of being hounded by questions that the client does not want to answer because it is hard to address those aspects of him or herself.

This remedy is good to detox vaccines particularly where someone has incarnated with blocked energy in their nervous system, exhibiting e.g. childhood precociousness and children who behave like teenagers from a very early age

OTHER

- Works well with Vaccinotoxin
- Works well with heart remedies: Cactus, Crataegus, and Digitalis

15. *Lasiocampa Quercus*

LASIOCAMPA QUERCUS

Oak Eggar Moth
Country: France
Order : Lepidoptera
Suborder: Moths
Family Lasiocampidae
Genus: Lasiocampa

BACKGROUND

The oak-eggar moth, a large moth with up to 90mm wingspan, is named because of its oak-acorn like chrysalis rather than because it feeds on oak trees. This gives the remedy a keynote of confusion and things being misnamed or misunderstood.

Oak egg moths are found in different habitats in Europe and are very much darker in the northern parts and at a higher altitude than in warmer climates and the south, this enables them to take in more heat from the sun. The female is much bigger than the male, with larger abdomen and bigger wings, is lighter in colour and flies at night and displays the usual moth behaviour of hovering and being attracted to flames. The male is smaller, darker red, brown in colour, with feathery antennae and flies by day. In the remedy picture this can be seen as shift work – irregular working hours with different partners going to work at different times. Males like the sun and are more active the sunnier it gets. The oak eggar moth caterpillars live communally, they are short with brown marks, with tufts of white hair on the sides of segments and a black line between segments. They live on hawthorn, leather, bramble, bilberry, broom, salix and sea buckthorn.

They have a two-year life cycle, with caterpillars hatching from May to September one year, then pupating the following May or June and emerging as an adult moth the following year.

The moths have two usual characteristics, they lack the usual wing-coupling devices of most moths (which aid hovering and can slow down flight reducing the number of wing beats to less than fifty per second) and they also have very much reduced tongues (which affects feeding).

KEY FEATURES

- Confusion, misunderstanding, people or things being misnamed
- Females with large stomachs
- Gastric disorders, bloating
- Children from boarding schools
- Shift work
- Males who work in outdoor jobs
- Frontal headaches, sharp pain worse for moving head
- Seasonal Affective Disorder (SAD)
- Skin problems due to sunlight
- Depression

PHYSICAL EFFECTS

- Frontal headaches – can feel constricted band-like tightening
- Sharp pain in temple particularly on moving head
- Eye strain
- Disorientation – feeling woozy as though can't hold head up
- Seasonal Affective Disorder, SAD – particularly in women
- Skin problems – mainly to do with stomach, indigestion, excess acid production and irritable bowel syndrome where there is a picture of undigested food or bloating
- Pituitary gland disorders

MENTAL / EMOTIONAL EFFECTS

- Females tend to be drab and blend into the background, don't want to be seen or make a fuss, often they are restless and can't sit down, fidget etc
- Males are attracted to outdoor jobs or like sunlight, they are prone to depression and can get into very deep, dark depressive states, are earthbound
- SAD depression and fatigue especially in the winter – affected by lack of sunlight
- Like to travel to different places
- Scared of flying
- Deception – there is often an aspect of deception with this remedy, people or things being misnamed or not being quite as they seem, when questioned they can become shifty
- Communal living – boarding school education
- Shift work – with partners not seeing each other for long periods of time
- Restlessness of a male – tends to flit from one thing to the next, not to settle

EFFECTS ON CHAKRAS / SUBTLE BODIES

- Blocked crown chakra
- Blocked solar plexus chakra
- Males are very earthbound
- Subtle bodies misaligned with subsequent disruption of internal energy and feelings of disorientation

SPIRITUAL ASPECTS

Oak eggar moths show the spiritual characteristics of moths and are particularly grounded into the earth – particularly the males sometimes lose their cosmic connection through blocks in their crown chakra. Then they can become very depressed with dark moods.

The females flying by night, the males flying by day signifies that in this remedy the two sexes have different time scales and can be 'out of synch' with each other. Males love the outdoors and sunlight, females less so. Both can be affected by SAD in the winter and need sunlight.

The females often have large stomachs and are bloated because there is something that they can't let go of. In both sexes there can be digestive problems, things are hard to deal with which brings about a state of nervous restlessness with energy jammed in the nervous system and an inability to settle. This is often related to deception, things not being as they seem.

OTHER

- Compare with Lithium, Nux vomica, goes well with Sol.

16. *Chorthipus Scalaris*

CHORTHIPUS SCALARIS

Grasshopper (male)
Country: France
Order: Orthoptera
Suborder: Caelifera (grasshoppers, crickets, katydids)
Super family: Acridoidea (grasshoppers)
Family : Chorthipus

BACKGROUND

Grasshoppers live in the grasslands and eat grass and some weeds. They often characterized by their 'song', which is produced by solitary males wanting to attract females using stridulated pads on their hind femurs to rub onto an ordered vein on their forewing. The song has a slow build up of short croaks getting quicker and louder until constant.

A note can last up to ten seconds punctuated by shorter, high-pitched notes. Song increases with the temperature.

Grasshoppers of the species Chorthipus scalaris are found in a variety of colours from yellow to green to brown. This specimen had red brown legs and was brown in colour. Camouflage is characteristic of grasshoppers who are believed to change colour to suit their environment. Usually they are solitary, but sometimes due to lack of food and changes in weather condition, they swarm together and become pests of crops. This does not happen as regularly with most grasshoppers as it does with locusts (which are a type of grasshopper). It has been found that locusts change colour to a dark brown. Their muscle tone improves as the weather becomes hotter and drier and there is a lack of food in the area, they swarm together. It had been found that 'serotonin' increases so that there is three times as much in 'gregarious' (swarming) locusts as in solitary ones. It has been suggested the same is true, to a lesser extent, of all grasshoppers.

Grasshoppers can fly, but use their legs to leap. When disturbed they can leap long distances and to heights of up to six feet in the air. They can release noxious chemicals when they are threatened.

Male grasshoppers are often larger than females. After mating, the female lays eggs in a foamy substance in mud, in plant roots and in manure. The eggs develop over winter until nymphs, which are immature forms of grasshopper, hatch. They moult four times before reaching adult size. The nymphs (unlike larval forms of other insects) are similar in form to adults and consume the same food. There is no dramatic metamorphosis as seen with butterflies, jus a gradual growth with moults. The stage from nymph to end of adult life is often about three months, so three quarters of their lives are inside the egg.

Grasshoppers often have problems with their urinary tracts and urinate continuously, especially when hopping around. They also spray faeces when they are getting ready to mate. Their oxygen is not carried in blood, which is green. It is thought that longer bodied forms of grasshoppers found in fossils from the late Permian period related to a time in earth's history when there was more oxygen in the air.

Grasshoppers are an important source of protein in Asia and are regarded as a delicacy.

KEY FEATURES

- Urogenital disorders – particularly incontinence
- Lung problems – particularly those resulting in lack of oxygen in the blood
- Diarrhea with loose stools which cannot be controlled
- Aftermath of cigarette addiction / ecstasy use
- Post operative – aids recovery especially after amputations or joint or organ replacements
- Depression / loneliness
- Fear
- Pent up energy
- Boys and men with autistic spectrum disorder
- Lonely people with a lack of connection to others
- People who walk away from confrontation and turn their energy inwards

PHYSICAL EFFECTS

- Chest – difficulty breathing, especially the inhalation phase
- Lungs – disorders resulting in a lack of oxygen in the blood
- Post operative – aids recovery especially after amputations or joint or organ replacement
- Urogenital disorder – particularly when they result in incontinence
- Diarrhea – with loose or watery stools and no warning so there is a lack of anal control
- Inflammatory diseases – e.g. arthritis
- Symptoms change with weather and temperature (better for heat, sun, summer)
- Symptoms improve with company and with lack of food
- Liver disease
- Kidney disease
- Clumsy people

- Comfort eaters
- Vaginal discharge – foamy in nature (note this can indicate Trichomonas infection)

MENTAL/EMOTIONAL EFFECTS

- The remedy is useful in the aftermath of drug addiction and follows on from remedies of drugs used tautopathically in potency, such as Opium 30c or other detoxing remedies, this insect remedy is useful after the 'withdrawal' phase
- Fear – particularly at night, fear of the dark, fear of specific things
- Lonely people – especially males who find it hard to connect to others, males with autistic spectrum
- Repetitive behaviour – it is difficult to move on the new things
- Camouflage – don't want to be seen
- Clumsy people who don't think before they act and can upset others unintentionally
- People who move away rather than confront things
- People who are moved by soul music
- The remedy can bring a sense of where to go next, a sense of direction
- Let go of things indiscriminately (even things they need to keep hold of)
- Can feel vulnerable – worry about what others think

EFFECTS ON SUBTLE BODIES / CHAKRAS

- Base and sacral chakras
- Energy disrupted in these two chakras resulting in indiscriminate letting go both physically and emotionally
- Astral body disrupted – lack of flow, sometimes it leads to deception because too many things have been released but a block in the lung area has not allowed the new impulses, this results in repetitive behaviour and a feeling of not knowing where to go next
- Kidney and lung chi are weakened
- There is often a block in the crown chakra
- Lack of connection to soul, higher self or spirit
- Brings back boundaries and heals tears in the etheric body

SPIRITUAL ASPECTS

- Opens the chest and depends on the breathing to receive new codes and impulses from the cosmos and higher self – especially through sound and light waves
- Balances letting go and incoming energies – can restore rhythm in their body
- Helps stabilize the sound waves in the body especially in the liver
- Sound healing can be helpful to use in conjunction with the remedy, this can help restructuring and reorganizing DNA and RNA within the cells and can open up new neural patterns
- Aids shamanic soul retrieval
- Helps to link to power animal
- Helps link to other members of the soul group
- People who need the remedy try and fill their lives either with activities or by comfort eating
- Enhances ones ability to connect across species and with aliens – may bring in alien or intergalactic ideas

OTHER

- Useful after remedies to help drug addicts quit – after the withdrawal phase
- Works well with other lung remedies
- Syphilitic miasm
- Compare with bowel nosodes
- Works well with aquamarine
- Enhances the properties of starlight elixirs and star remedies

17. *Schitocerca Gregaria*

SCHISTOCERCA GREGARIA

Desert Locust (male)
Order: Orthopptera
Sub order: Caelifera (grasshoppers, crickets, katydids)
Family: Acrididae (grasshoppers)
Genus: Schistocerca

BACKGROUND

The provings for Schistocerca Gregaria, migratory locust were found to be very similar to those of grasshopper, with lung problems, urogenital disorders, particularly related to incontinence and diarrhea with loose stools. In the locust there is little that can be done to stop bowel leakage, people who have to go now and risk soiling their clothes. In the locust, there was more emphasis on food and diet – either the need to eat or to starve, so it is useful in any form of eating disorder or with obesity issues resulting from an urge to eat, particularly people who snack a lot. One prover talked about an irresistible urge to eat in the middle of the night, others about feeling they must eat in case it was 'their last meal', or the need to starve and make sure she kept the food for when she really needed it.

However, in locusts depression and self loathing played a strong part of the mental picture. Often there were issues of self esteem and boundaries with other people and a lack of self confidence which either came along with solitary behaviour (loners or people with few friends, but as one prover commented they will claim to have thousands of friends on Facebook!) The other side of the remedy is in group behaviour as this is a key remedy for youth gang culture, often with yobbish behaviour. People feel happy being camouflaged by the anonymity of being part of a gang and the respect that being one of the gang engenders. Here 'fitting in' is important camouflage for the client, they willingly sacrifice large parts of themselves to look credible in the eyes of their peers and to gain respect. Anger, violence, alcohol and drug use often play a part in the picture. The musical aspects of grasshopper remedy are not as important in locusts, although music, particularly rap and street music may be part of what is needed to fit in to the gang.

Grasshoppers and locusts are physiologically and anatomically very similar, hence the similarities of the remedy pictures.

The key difference between locust behaviour and that of grasshoppers is their ability to respond to changes in conditions by leaving a solitary lifestyle to gathering together with many other locusts to form swarms which can result in large scale crop losses. The desert locust is one of the world's most dangerous pests because of its ability to swarm and fly rapidly across large distances. In the gregarious (swarming) phase there can be as many as five generations in one year. Desert locust plagues have threatened agricultural production in Africa, the Middle East and Asia since ancient times.

Desert locusts live a solitary life until it rains. They behave like solitary grasshoppers, living on their own, eating vegetation and fruit – only coming together to mate with others. They lay their eggs 4 inches below the surface of the soil in a frothy compound which dries into an egg pod holding 80-100 eggs. These hatch after between 10 days in hot climates to 70 days in colder climates as flightless nymphs called hoppers. The hoppers are very vulnerable, go through several moults and mature into winged, flying adults in 30-50 days. They disperse on their own and are well camouflaged, being generally coloured green or brown.

The rain causes new vegetation growth and softens the sand allowing the female locusts to lay their eggs in the ground more easily. The new vegetation provides food and cover for the nymphs as they grow into adults. These hoppers have to congregate together to feed and in doing so bump into each other and the close physical contact particularly between their hind legs stimulate metabolic and behavioural changes. They thereby transform from the 'solitary' form to the 'gregarious' form. Their skin changes colour from green to yellow and black, the adults change from brown to red (immature) or yellow (mature). Their bodies become shorter and they give off a pheromone that causes them to be attracted to each other, ultimately leading to swarming. The rate of sexual maturation increases dramatically with overcrowding. The chemical changes stimulate the females to lay more eggs under swarming conditions and for the whole life cycle to speed up so there are more generations in a shorter time.

Locust swarms fly and can cover 100-200 miles per day as high as 2000 feet above sea level. Each locust can consume the approximate equivalent of their own body mass (2g) per day and they eat a great variety of food and agricultural crops. Their droppings are toxic and spoil any left over uneaten or stored food. A single swarm can cover 1200 square kilometres and contain between 40 and 80 million locusts per square kilometre. Locusts can live for three to six months and there is a ten to sixteen fold increase in locust numbers from one generation to the next during the 'gregarious' or swarming phase.

Desert locusts have proved difficult to control partially because of their ability to resist and adapt to certain chemicals but mainly because the plagues often occur in remote areas of underdeveloped countries, where there is insufficient infrastructure to deal with the scale of the problem and the fact that once swarming, locusts can travel large distances very quickly. In 1985, a plague of locusts started in Ethiopia and Sudan, travelled across to the West Coast of Africa then across the Atlantic Ocean to Barbados 4800km away.

It has been estimated that if locusts in a swarm covering two square miles reproduced over four generations (between a year and eighteen months) there would be a severe threat and infestation of the surface of the Earth, spanning 196 million square miles. Fortunately, changing climatic conditions (e.g. drier weather), insect pests along with other predators such as ants and birds, and insecticides stop the scenario by triggering the locusts to revert back to solitary behaviour. Examples of locust insect pests include the stomorhina fly, which lays its eggs on the locust pods, its grubs then eat the locust eggs before hatching. Also the Trox beetle larvae which destroys locust egg sites entirely. In terms of insecticide control, one light aircraft carrying 60 gallons of insecticide can kill 180 million locusts.

Swarming behaviour is governed by chemical changes in the locusts. There are key changes in the pheromones given off by the adults and by the hoppers which promote the behaviour. Recent research points to a serotonin-like neurochemical being responsible for the changes. Serotonin in humans plays a key role in well-being and is lacking in the brain of patients with primary or endogenous depression. Again, depression and loneliness come into the remedy picture more in locust than in grasshoppers. The serine protease inhibitor greglin also plays a part in the process, it inhibits serine proteases and enzymes of elastase and chymotrypsin which play a role in changes in connective tissues (and in inflammatory diseases and cystic fibrosis in humans).

Interestingly, whilst all three Caelifera remedies: grasshopper, locust and cricket, all feature 'pent up energy' and sometimes a need for quick sexual release, locust seems to show more explosive outbursts of anger than the other remedy pictures. This fits with the 'gang mentality' which is the keynote of the desert locust. Whilst all three show signs of irritation, this is more prevalent in both cricket and locust rather than in grasshopper. Grasshopper people are much more likely to let things go and walk away from confrontation and turn their anger in on themselves than locusts who tend to be confrontational.

Fear is also a feature in all three remedies, sometimes showing as night terrors in children, particularly in locust and grasshopper. Fear of what other people think of them or fixation that someone is 'out to get them' intensifies in locust. Fear of the dark or being buried alive have shown in the provings of grasshopper and locust.

KEY FEATURES

- Gang culture – this is a keynote of the locust remedy, people who are drawn to gangs, camouflaging themselves within the anonymity of the group, they have an urge to fit in and respond to peer pressure, often losing part of themselves in doing that, explosive violence, fighting, heavy drinking, alcohol, drug , cigarette and ecstasy usage and the need to gain respect on the streets at whatever the cost to ones' self
- Depression
- Self esteem issues, issues with boundaries with others, lack of self confidence
- Coping with changing conditions
- Eating disorders – need to eat / need to starve / obesity with snacking
- Fear – of specific things, fear of the dark, fear of being buried alive, children with night terrors
- Urogenital disorders – particularly those resulting in incontinence
- Lung problems – particularly those resulting in lack of oxygen in the blood
- Diarrhea with loose stools which cannot be controlled
- Inflammatory diseases
- Post operative – aids recovery especially after amputations or joint or organ replacements
- Boys and men with autistic spectrum disorder
- Lonely people with a lack of connection to others, (but may have lots of 'friends' on Facebook)

PHYSICAL EFFECTS

- Chest – difficulty breathing – especially the inhalation phase
- Lungs – disorders resulting in a lack of oxygen in the blood
- Eating disorders – people show a need to eat or to starve and 'conserve' food, this can range from obese people who snack a lot, or need a particular food at a particular time of day to the extremes of anorexia and bulimia, but food often is a feature of the remedy picture in some way
- Inflammatory diseases
- Post operative – aids recovery especially after amputations or joint or organ replacement
- Urogenital disorder – particularly when they result in incontinence
- Diarrhea – with loose or watery stools and no warning so there is a lack of anal control
- Symptoms change with weather and temperature (better for heat, sun, summer)
- Symptoms improve with company and with lack of food
- Liver disease
- Kidney disease
- Clumsy people

MENTAL / EMOTIONAL EFFECTS

- People who change approaches, appearances and points of view a lot
- Response to changing conditions in life
- Chameleon-like people
- Comfortable in groups – like to fit in, often losing something of themselves along the way
- Power, control, respect issues
- Gang members / hoodies
- Explosive angry outbursts, extreme irritation, sometimes leading to violence and fighting
- Fear – of specific things, fear of the dark, fear of being buried alive, sometimes they are fearful for no reason, sometimes the fear can focus onto someone they perceive as being 'out to get them'
- Children with night terrors

- The remedy is useful in the aftermath of drug addiction and follows on from remedies of drugs themselves or opium and detoxing remedies, useful after the 'withdrawal' phase
- Lonely people – especially males who find it hard to connect to others, autistic spectrum, sometimes they hide this eg. by saying they have lots of friends on Facebook
- Repetitive behavior – it is difficult to move on to new things
- Camouflage – don't want to be seen
- Depression or loneliness
- Clumsy people who don't think before they act and can upset others unintentionally
- Let go of things indiscriminately (even things they need to keep hold of)
- Can feel vulnerable – worry about what others think

EFFECTS ON SUBTLE BODIES / CHAKRAS

- Throat chakra – heart chakra disruption, heart chakra often partially blocked or too wide open which leads to either a lack of connection to the world and / or other people or too much influence on his life by someone else, this can result in power struggles or sacrifice
- Astral body disrupted – lack of flow, sometimes it leads to deception because too many things have been released but a block in the lung area has not allowed the new impulses and energy in, this results in repetitive behavior and a feeling of not knowing where to go next
- Kidney and lung chi weakened
- Lack of connection to soul, higher self or spirit
- Brings back boundaries and heals tears in the etheric body

SPIRITUAL ASPECTS

- Opens the chest and depends on the breathing to receive new codes and impulses from the cosmos and higher self, especially through sound and light waves
- Atlantean karma – power struggles and or controlling or victim behaviour patterns
- Helps link to other members of the soul group
- People who need the remedy try and fill their lives either with activities or by comfort eating

- Enhances ones ability to connect across species and with aliens – may bring in alien or intergalactic guides

OTHER

- Useful after remedies to help drug addicts quit – after the withdrawal phase
- Works well with other lung remedies
- Compare with bowel nosodes
- Works well after Berlin Wall for power struggles
- Syphilitic remedy
- Often symptoms worse in damp and wet and heat and at night, better for cold, better for being alone, intensified in crowds

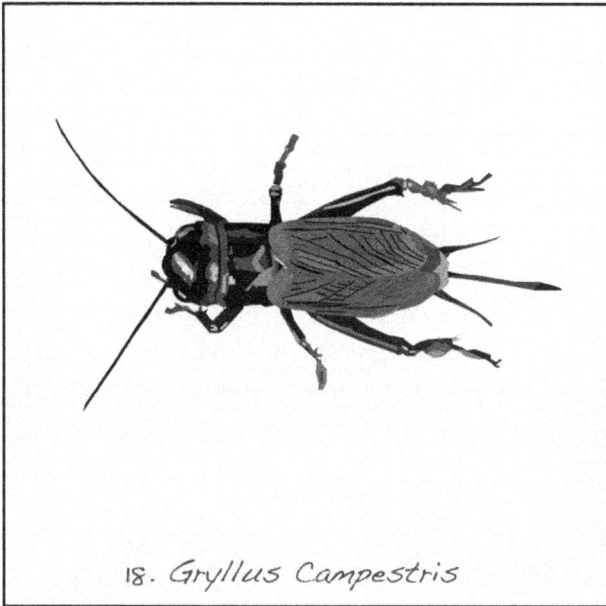

18. *Gryllus Campestris*

GRYLLUS CAMPESTRIS

Field Cricket (male)
Country: France
Order: Orthoptera,
Suborder: Enfifena
Super family: Grylloidea
Family: Gryllidea (common crickets)

BACKGROUND

Crickets are related to grasshoppers, bush crickets resemble them in appearance, but true crickets have flatter bodies and long antennae. They are nocturnal, and like grasshoppers have strong, jumping hind legs. However, their front legs are adapted for digging. They hold their forewings horizontally over their bodies, which are flatter than other crickets. As in all members of the order, orthoptera – their ears or membranes are found on their forelegs rather than their heads (in grasshoppers the ears are placed on the abdomen).

Male crickets produce as song a chirp, by raising their left forewing to a 45 degree angle, and rubbing it against the upper hind edge of the right forewing which has a thick scraper. This action is called stridulation. The song is specific. There are four types of cricket song: a loud calling song to attract females and repel other males, a quieter courting song to a female, a copulatory song after mating and an aggressive song triggered by chemoreceptors on the antennae stimulated by the presence of another male cricket or a predator. Cricket song is usually higher pitched than grasshopper song. The rhyme and pulse of song is important. A single, solitary male will sing and other males will join in and harmonise. They don't sing in the rain and rarely sing on cool or windy nights.

Different crickets chirp at different rates depending on their species and on the temperature (crickets chirp at greater rates with increasing temperature). In the snowy tree cricket from the USA, Dolbear's Law makes it possible to calculate the temperature in Fahrenheit by adding 40 to the number of chirps produced in 15 seconds.

Crickets attract parasites, particularly the tachinid fly ormia ochracea which uses cricket song to locate a male and deposit her larvae on him – this has resulted in some crickets maturing to be unable to chirp in Kauai, Hawaii.

Crickets are omnivorous and feed on organic material, decaying plant material, fungi and seedling plants. The have even been known to eat other dead crickets and become predatory to weak members of their own species when there is no other food available. Crickets have very powerful jaws and have been known to bite humans when picked up. Nymphs live mainly on plant roots.

Crickets mate in the late summer, the females lay as many as 2,000 eggs in the autumn, the eggs are laid in burrows in the ground near plant roots. They are laid through an ovipositor, which are prominent in female crickets. Female crickets often visit where they have laid their eggs. The nymphs undergo approximately ten moults before they reach adult size.

KEY FEATURES

- This is a right-sided remedy – symptoms are better at night, worse in damp, windy or rainy weather. Often there is some degree of fatigue as a concomitant or associated feature with other symptoms.
- Parasites
- Pent up energy
- People who don't quite fit in
- Irritation – suddenly there is a sense of irritation with something that has never been seen as irritating before, the irritation intensifies
- Lives can lack a sense of structure and purpose and have little rhythm to them or have a sense of irregular rhythm, so there are periods of almost frenetic activity followed by periods of little action
- Lung disease, ribcage or rib pathology
- Intestinal problems, including anorexia
- Successful businessmen
- Itchy genitalia in males
- Thrombosis
- Arrhythmia
- Women with ancestral problems, patterns running through the female line
- In females there can be a desire to remain asexual with delayed onset menstruation, flat chested women who are thin are the characteristic of 'cricket' types
- Creative people with lack of music in their life
- People who have difficulty telling right from left or who have difficulty with directions

PHYSICAL EFFECTS:

- Right-sided headache, constantly aching, wearing you down
- Pent up energy
- Lungs – this is helpful for all lung problems, particularly with constricted airways, wheezing and coughing, pleurisy, emphysema, asbestosis, tracheostomy, pleural taps, or drainage, pneumothorax

- Works well with ant remedies (Formica) particularly with Comahen
- Irregular breathing
- Skeletal Problems – either stiffness without pain, particularly in the right side of skeleton or upper back, neuralgia with shooting pains radiating out from the bone
- Rib pathology
- Curvature of the spine
- Irregular heartbeat – arrhythmia and tachycardia
- Thrombosis
- Intestinal problems with belching and flatulence, anorexia
- Constipation – dry stools which scratch on evacuation
- Pent up sexual energy, sometimes need for quick release
- Itchy genitalia in males
- Delayed onset menstruation in females
- Sensitive to sounds especially high pitched sounds

MENTAL / EMOTIONAL EFFECTS:

- People who don't quite fit in
- Males often get a very strong rush of forceful, dominant male energy towards the end of puberty and get very driven. They won't take no for an answer. This can lead to them being forceful towards women or having an urge to hit out if people don't do what they want. Older males are often successful businessmen who forego relationships to build their successful careers – but at some later stage they question this and feel lonely, often not knowing how to start relationships at this stage of their life.
- Sometimes males are on the autistic spectrum and can have 'one tracked' almost obsessive-compulsive behaviour e.g. they are avid collectors or have a passion for complex mathematical problems etc
- People who appear hard on the outside and unemotional, the client case may be difficult to crack
- Females are often quieter and take on other people's problems and issues even if they have nothing to do with them. This is especially true of 'ancestral' problems. They carry 'the sins of the fathers' and patterns from several generations back. There may also be family issues within living memory which affect the case. Often women try to be peacemakers and compromise themselves for the sake of others being happy

- Urge to scream – so someone will notice
- Food is important and features in some way – either someone concerned with nutrition, or food allergies, diet or weight issues or have particular food preferences
- Cross gender confusion in both sexes
- People who have difficulty telling right from left or poor sense of direction
- In females there can be a desire to remain asexual with delayed onset menstruation problems with anorexia etc, flat chested women who are thin are the characteristic of 'cricket' types
- Creative people with lack of music in their life
- Lives can lack a sense of structure and purpose and have little rhythm to them or have a sense of irregular rhythm, there are periods of almost frenetic activity followed by periods of little action
- Irritation – suddenly there is a sense of irritation with something that has never been seen as irritating before, this irritation intensifies and brings a sense of unease

EFFECTS ON SUBTLE BODIES AND CHAKRAS

- The main effect of this remedy is to clean and unblock the energy channels in the bodies. It helps to align the chakras. The throat chakra is unblocked and the remedy can help restore communication or aid one to say what needs saying.
- The nadiis are cleaned, particularly if used in conjunction with rose or rose oil or essence or deodar (Himalayan cedar)

SPIRITUAL ASPECTS

- Brings one into the present in a grounded way
- Brings acute hearing – clairaudience
- Restores rhythm in the breathing system and circulatory system, helps to regulate the heartbeat
- The remedy helps clear ancestral patterns particularly those that have been carried 'in the blood', it can be useful for ancestral feuds, curses and intertribal/interfamily disputes
- There is a strong link to 'Adam Kadmon' – the ideal human blueprint, which neither male or female, so there can be gender confusion

issues, (the Elohim breathed life into Adam and the Jehovah pulled out a rib to make Eve and form the separate sexes)

- There is a link to a time in human evolution before the lungs were fully formed, when gaseous exchange root place through the pores in the skin
- The remedy treats interior wind states especially where there is some depletion of chi or liver/blood yin deficiency, this helps to treat neuralgia, stiffness and cramp
- The lung-colon paired relationship stimulates and regulates wind flow through the intestines
- As this species of cricket has refined directional sensitivity, remedies can help give a sense of direction both in terms of life purpose and in terms of direction e.g. left / right, compass directions etc, it can help one find ones way both esoterically and exoterically
- As with many insects remedies there is a sense of irritation, which in this remedy picture comes from something which previously had not been a problem but then starts to irritate, and irritation grows. Irritation produces 'imperil' where there is astral body agitation of the nervous system can lead to serious psychological problems.

OTHER

- Compare to Stannum
- Works well with rose either as homeopathic, essence or oil
- Works well with deodar (Himalayan cedar)
- Complements Formica

19. *Dynastes Hercules (Male)*

DYNASTES HERCULES (MALE)

Hercules Beetle
Country: Iquitos, Peru
Class: Insecta,
Order: Coleoptera,
Family: Scarabaeidae,
Sub family: Dynastinae,
Genus: Dynastes,

BACKGROUND

Dynastes Hercules is a large, scarab beetle of the Rhinoceros Beetle type, so named because of the large thoracic horn in the male. It also has a horn coming out of its head below the thoracic horn. It has thick hairs on the thoracic horn. The Electra is waxy, and was reddish black with a black head, the colour of the Electra changes to black as it dries out and the reddish black as it rehydrates.

Hercules beetles are generally 110mm (4.5ins) long and are generally prized by insect collectors and have been used as talisman by indigenous tribe's people in the rainforest. They are in relative size the strongest animals on the planet as they can carry 850x their own body weight (average weight of a Hercules beetle is 1oz, they can lift an equivalent of 53.125 pounds in weight over their head – the equivalent of an average sized man lifting a battle tank).

Hercules Beetles are generally nocturnal and non-aggressive, they live on the floor of the rainforest, becoming more active with heat and humidity.

There is a sexual dimorphism, in that females do not have the horns of the males, and so may be perceived as being smaller, whereas their actual body size is not.

In the rainy season, the males become increasingly aggressive towards other males and can become territorial. There is a great deal of fighting between the males with the horns and spiky protuberances on the legs being used to lift the other male up and throw him onto the ground.

The Hercules beetles have a lifestyle which undergoes a complete metamorphosis. Mating takes place and the female lays eggs directly into the soil when the weather is warm, usually during the rainy season. The eggs hatch into larvae, after about a month, which look like hairless caterpillars with chewing mouthparts. There are three instars (larval stages) after each the larvae moult. The final instar leave larvae of up to 110mm long weighing 120g. The larvae live underground, feeding on decaying plant material, logs, tree stumps and dead leaves. They sometimes tunnel up from the ground into rotting tree stumps. They live in the larval stage from six months to two years. They pupate under ground and the adults emerge after 6-8 weeks, living under the soil until spring. The beetles live for only 3 months to a year as adults. As temperature and humidity rise it triggers the females to release pheromones and to mate. The height of release of pheromones from the females usually coincides with onset of spring or the rainy season. The males then become much more aggressive and fight the other males. Mating takes place and the females lay their eggs directly into the ground.

Hercules beetles are generally nocturnal and non-aggressive, they hide when they hear other animals. The adults live on the rainforest floor feeding on rotting fruit, tree sap and decaying plant material.

The larvae are a food source for small mammals and predatory soil-dwelling arthropods e.g. centipedes, ground beetles and some spiders. They are vulnerable as they do not have the exoskeleton of the adults. The eggs are often eaten by a mite (which is the same size as them), also the mydas fly maggot, some ants and bees.

KEY FEATURES:

- Male sexual issues and dysfunction
- Bravado, boasting, posturing, peer pressure, intimidation – needs to be seen to be the best
- Warrior types from yobs to private generals
- Brings strength and courage
- Spinal problems
- Being stuck in the shadow of one's teachers – not having the courage to stand-alone and learn from one's own mistakes.
- Excess spiritual zeal leading to not being able to function in the world
- People on a spiritual journey working towards integration of soul and personality
- Crises in life brought about by experiences in the material realm, but ultimately to help you develop spiritually by learning life lessons
- Dealing with issues of mass consciousness, media hype, illusion
- People who cannot function properly in the world or who have given everything up to develop spiritually

PHYSICAL EFFECTS:

- Helps people with 'rugby player' physique – strong shoulders and legs
- Occipital headache – can be accompanied by feeling 'not quite here' or feeling of expansion or denseness, headache worse on movement, better for staying still, pain is intermittent

- Stiffness in back of skull and neck area
- Sore throats (bacterial infections)
- Damaged vocal cords and larynx – useful support for artificial voice boxes e.g. after cancer
- Misalignment of shoulders, one higher than the other
- Spinal problems and misalignments – the remedy helps align the vertebrae and strengthens the spinal column, osteoporosis, collapse of vertebrae, aids physical atlas bone adjustment
- Loosens up tension in muscle skeleton system
- Displaced joints
- Ligament muscle and tendon problems, supports chi kung style bone-breathing meditations and a physical regime e.g. exercise or osteopathy
- Muscular degeneration
- Lungs – aids the inhalation phase
- Motor neurone disease and other degenerative diseases of the nervous system – this is a very helpful remedy for any pain of nervous origin
- Liver / gall bladder disease
- Spleen disorders
- Male sexual dysfunction – the remedy shows a picture of extreme male sexual energy, over-focus on sex which takes over everything else, males exhibiting boasting and bravado, wants it to be known that he can have sex with anyone he wants (usually the most desirable women), pattern of one-night stands and imposing his will on women
- The remedy can be used for impotence, ejaculatory problems, injury to the male reproductive system, Peyronies disease or excess of testosterone, brings the male sexual energy back into balance.
- It appears to be less useful as a female remedy, e.g. in vitro fertilisation, genetic disorders or inherited patterns
- Recuperation – after illness or long stays in bed, helps to rebuild and tone muscles, bring back vital energy
- Chronic fatigue syndrome
- In most cases the symptoms are better for heat and damp and aggravated at night

MENTAL AND EMOTIONAL EFFECTS:

There is a gradation of people for whom the 'Hercules' remedy might be useful as it is helpful for those on a spiritual path, at whatever stage they are on the journey. Therefore, there is a range of mental and emotional states which can apply:

- Successful people, particularly men, who force their will on others and 'know what is best'
- Can be vain, people who over-celebrate their successes (can be a 'party' animal)
- Lectures to others and with body language but doesn't take advice from others
- Knows how to deal with failure, can turn mistakes into successes
- Leads from a feeling of superiority, feels better than others
- Doesn't suffer fools gladly
- Sadness, melancholy, loneliness of the warrior – standing alone with a heavy heart, feeling misunderstood
- Need for success to recognised by others
- People who have difficulty to accept, learn from or move on from failures or mistakes, either by not engaging with life for fear of failure or repeating the same mistakes
- Being stuck in the shadows of your teacher – people who look for the next workshop or latest book or see their guru rather than standing alone and putting into practice what they have been taught, this often relates to willingness to learn from their own mistakes rather than the wisdom of the teacher
- All warrior archetypes
- Yob culture, hoodies, football hooligans, calling people out, becoming violent to show off to others – challenging people
- Female passivity
- Male sexual bravado – likes to boast about conquests, lots of one-night stands, can get any women he wants
- Lack of connection to sexual partner
- Promiscuity
- Sexual dysfunction with mental origin
- Can lose all mental focus because of excess sexual energy – thinks about sex at inappropriate times

(the sexual aspects of the remedy are more related to males than females)

- Claustrophobia – helps housebound people expand within the world
- People who are swayed by the world or mass emotion e.g. grief, fear, anger
- People who are swayed by media hype – need to stay young, issues of war and peace etc
- Dealing with duality – the remedy helps one to develop a sense of oneness with everything around you
- Dealing with desire – desire for material things, money, success, nice house, good car etc
- Fear – can help bring courage and help one achieve goals and targets
- Anger – can help the client deal with anger in a different way, both bringing peace and the use of anger appropriately
- Detachment – some people have become too detached from the world and people around them so they disengage from the material world and are looking for a 'spiritual' answer for everything
- Ungroundedness – giving everything up for spiritual advancement, people who have difficulty functioning in the world
- Cases of excess spiritual zeal that lead to fanaticism and even insanity around the individual
- Can help when a person is at a cross roads and needs to choose between the easy but fulfilling path or turbulent but fulfilling one.
- Dealing with crises in the material world which are showing the individual their 'life lessons'
- The remedy helps one deal with the issues that come up as the soul impulses become integrated with and often overcome the desires of the lower personality. It brings the ability to see through the glamour and illusion of material world. It helps to develop tranquility, courage, discrimination, right speech and the need to serve the group (even doing menial tasks when needed). It brings a quest for truth bringing discernment and discrimination. It helps to master the lower desire nature, sensation and good intentions which have not been carried out. It ultimately brings mastery of the mind and emotions.

EFFECTS ON SUBTLE BODIES, CHAKRAS:

- Solar plexus can be very uncomfortable when dealing with power issues
- Right breathing helps direct energy to the lower hara and sexual centres to help physical and sexual issues
- Third eye expansion related to a sense of knowing when to act
- Energy needs to be centred in hara to counter balance a sense of heaviness
- All chakra become balanced
- Antakharana – this is the spiritual channel of alignment along the vertebral column, it becomes clear with vital energy (electrical, magnetic, gravitational), there is an increase in light both in the Antakharana and throughout the subtle bodies
- Subtle bodies become balanced and harmonised
- Aura expands as do all the subtle bodies

SPIRITUAL ASPECTS

- Hercules is a remedy for those at any stage of the spiritual path, it helps them address the current issues and problems, to overcome crises in their lives, helps them find the solution, dealing with individual subtle body challenges, ultimately helping to align all the subtle bodies and bringing them together with the soul in atonement (AT-ONE-MENT)
- Excess use of will to manifest things without taking into account the needs of others (in extreme cases this can be shown as using black magic for own purpose)
- Commanding others but from the belief that you are superior than others
- Overcoming spiritual glamour and feelings that you are doing great work, healings etc, rather than realizing that healings happen when the client wants this, that you are just the facilitator
- Overcoming dependency on teachers, gurus, books or workshops instead of putting into practise what has been learnt and taking responsibility for ones own spiritual development
- The remedy helps expansion physically, in terms of the different subtle bodies but also spiritual expansion. This can be seen in

different ways, e.g. in terms of enhanced or quickened spiritual development, more balance between the spiritual and material worlds, subtle bodies integrating and coming into balance.

- Overcoming duality – the remedy brings in a sense of unity consciousness, an awareness of the interconnectedness of all things
- Overcoming issues of sex – moving from gratification of base needs of the personality through sex, to the sacred marriage – balance of feminine and masculine, spirit and integrated personality, and ultimately to greater creativity when sexual energy is raised from sacral centre
- The remedy enhances communication at all levels and can enhance spiritual capabilities e.g. clairvoyant, clairaudience etc
- If there are power issues then the solar plexus energy can be very unbalanced and can be felt as physical discomfort, this can be related to Atlantean karma
- Excess spiritual zeal leading to not being able to function in the material world
- Helps remove implants in brainstems
- Learns from his mistakes – preserves initial success
- Helps overcome life challenges which come from soul / personality clashes
- Helps one to stand alone and demonstrate mastery over thought and emotions and mastery over the material realm
- Helps one learn from experiences in the material realm which give you knowledge and understanding of oneself which is needed to help one serve others
- Help one become detached from personality-based desires, without becoming too detached from interactions with others
- The material world is seen for what it is
- Promotes mastery of the lower mind, overcoming fear, power, desire, sex, money, materialism etc which come from separation. Mastery of thought, restraint of speech and realisation of the truth behind things, brings in tranquillity, courage and discrimination
- Helps balance material and spiritual realm, dealing with issues of mass consciousness

OTHER

- Strong warrior spirit
- Leo astrological types with issues of power and control of others
- Altair and Aquila star remedies – regeneration right down to DNA/RNA or cellular level
- Links to sacred scarab beetle in Egyptian times

AND ONE SPIDER

"Whilst working with the insect remedies I was given a female golden orb spider which had been on display and eaten all the other spiders in the cage. This book would not be complete without this remedy".

Jenny Jones

THE ESOTERIC NATURE OF SPIDERS AND SPIDER REMEDIES

Spiders are animals that live from the perspective of spiritual world, with a sense of the interconnectedness of all things. In that sense, they are unlike insects and other animals which wish to 'escape the earth' – they are already living from a perspective that other animals aspire to.

In prehistory, there was a time of 'the Fall' where humans 'fell to earth' and became bound up in a materialistic, duality based illusion of life. This caused a perception of oneself as separate from everyone and everything around one. Humans 'forgot' the spiritual truth that everything and everyone is in interconnected and that everything they do affects the world around them. So most humans lost their links with the spiritual world as an underpinning of their earthly existence. This process of forgetting or falling was aided by the seeding of the earth by various different races of extraterrestrial origin.

Spiders did not fall to the earth and lose their connection to the spiritual world in the same way that humans and other mammals did. As spiders live from a spiritual perspective, but live on earth, they retain a strong cosmic connection and can bridge the gap between the spiritual and mundane worlds. They have thus found a way to live in a world based on duality, where popular opposites like 'good' and 'evil' give rise to judgements and a sense of separation and aloneness. However, they had retained the knowledge that behind the illusion of the material world is the spiritual world. This is seen as a continuum where external events have a purpose.

Spiders often bring up feelings of fear in humans and this can be related to the way that they live, 'entrapping' other animals by entangling them in webs. Spiders are very connected to webs of information that permeate through different dimensions and planes of reality. They therefore have strong links into 'duality based' binary computer technology and the 'world wide web' of the Internet through which the influence of ahrimanic forces can be felt. Many souls are now being trapped by this duality-based technology, into the confines of the World Wide Web and digital communications networks. Spiders however can go beyond this and link into other systems of non-binary information webs, much in the same way as crystal skulls do.

Some spiders also link into luciferic forces which bring light into the darkness of human consciousness and help people to remember their spiritual roots and interconnectedness with all things, aiding them to link into multi dimensional, unity based reality and redevelop links with the cosmos.

SPIDER REMEDIES, THE NERVOUS SYSTEM AND NATURE OF THOUGHT

The brain functions as a conductor of spiritual and cosmic thought and is the point of contact for spiritual will, the Atma. The nervous system is a network of energies and forces which are the outward expression of the inner, vital dynamic network of the etheric body and the millions of nadiis vary according to the personality ray of expression in any lifetime.

Human thought does not come from within the brain, instead thoughts project from the spiritual and cosmic realms and imprint upon the brain. The brain and nervous system can also operate through 'duality' where there are only two states of nerve activation – 'firing' or not 'firing' a fixed action potential along the nerve axon and triggering fixed chemical signals across the synapses. This makes the human nervous system and thought processes very susceptible to duality based binary/digital technology which creates an alternative reality of instantaneous 'heaven on earth'. This is particularly true for children and young adults who spend a lot of time with computers, digitalised TV or music and digital telecommunications networks. The semiconductor microchips used in these systems, whilst based on crystal technology are built to resemble spiders and signify the connection between spiders and this technology.

The motor nerves which originate from the cerebral cortex of the brain to travel down the spinal cord and reach the muscles, however also embody a 'sensory' aspect, sensing the movement of the body. This movement activity is fed back to the nervous system through the motor nerves as well as the sensory nerves, which convey information regarding the outside world from the sensory organs (eyes, ears, skin etc). Nonetheless, the sensory nerveds also have a 'motor' function in that they enable the astral body of the individual to reach towards it the event or activity that is being perceived.

The spider remedies link into the arachnoid mater or middle sheath/ meninges around the brain and spinal cord. The arachnoid mater even appears like a spider web under the microscope due to its fine networks of blood vessels. The meninges are part of a seventh ventricular space for cerebrospinal fluid. They enable thoughts from the human brain to be relayed back to the cosmos. Thus they partially function as a feed-back loop to the cosmos.

The nervous system in humans is often cluttered by various implants and devices to ensure that the illusion of separation from the spiritual realm is maintained. Spider remedies can help remove these and assist people to regain their spiritual connection.

Spider remedies also have a close link to the etheric body and the vital energy, with its anabolic building-up forces, these enable regeneration of the physical body and tissues. Spiders suck out all the liquidfied parts of their prey, leaving the structure dried out. These remedies can thus help people who feel that their etheric vital forces have been drained or sucked out of them. In terms of digital technology, this involves ahrimanic forces seeking out vital etheric energy to animate itself. This often either causes holes in the human aura. Spider remedies help to mend these holes and re-establish boundaries for the different subtle bodies, especially the etheric body and within the nervous system.

Within the physical body there are constellation and star maps of the cosmos and an indwelling memory of the precise position of the stars, including those where the client has spent previous incarnations. Between incarnations this map is transferred from the decomposing etheric body of the former life into the embryonic etheric body of the next, and this is also known as the etheric seed atom. Thus spider remedies can help link into those lifetimes and enable direct links to other star systems, which directly influence human evolution on earth.

Spiders are generally solitary creatures, spending a lot of their time being still and waiting for their prey to get caught in their web. Thus spider remedies are good to calm down overactivity, both physical or mental.

Usually spiders have pristine webs and remake them into different locations. So some spiders have a pattern of moving on and moving home, whilst others mend their webs in situ. Most spiders are precise and tidy in their web making and this is a feature in the remedy picture, particularly of women.

Spiders often generate fear in humans and the remedies are good to help those with general fears, such as fear of growing old, losing their looks, fear of what people think of them. They can be used with teenagers to address peer pressure. Fear is an emotion held in the kidney-adrenal system, thus spiders are often helpful to treat both physical and functional/stress related kidney or adrenal problems.

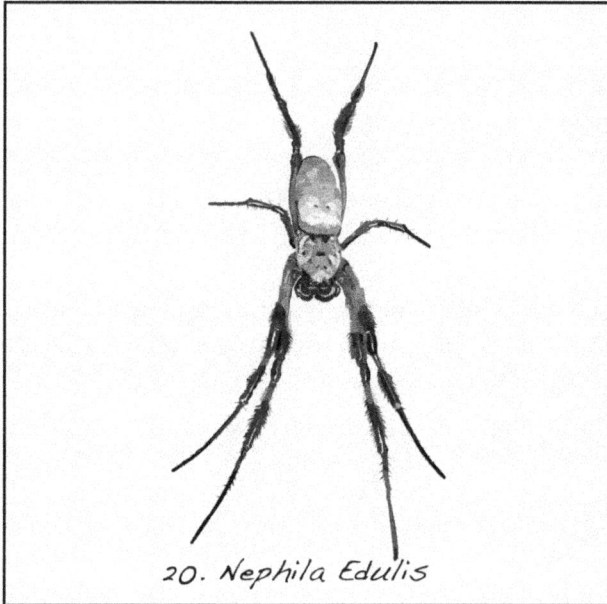

20. Nephila Edulis

NEPHILA EDULIS

Golden Orb Spider (female)
Country: Australia
Class: Arachnida
Order: Araneae
Suborder: Araneomorphae
Family: Nephilidae
Genus: Nephila

BACKGROUND

This female specimen had been kept on display in a cage with other golden orb spiders and had eaten them.

Golden orb spiders live in warm, moist areas – gardens, trees, and farmland in the tropics, and eastern Australia. Golden orb spiders were some of the earliest recorded web weavers and date from the Carboniferous and Permian periods.

Female golden orb spiders reach up to 45mm in body length, males are much smaller – sometimes 1000 times smaller. Normally female spiders are harmless, they allow other spiders and males to coexist symbiotically on their webs. They may bite if threatened, causing paralysis of their prey. Male spiders frequently creep up on the female, often when she is sleeping. Mating lasts up to fifteen hours. The female spins a sac for her eggs, digs a pit in the ground and buries it under the soil and debris. The spiders hatch with their egg yolks still attached, they stay together whilst their mouth parts, venom glands and spinning organs are immature, and disperse when they are fully grown. A lack of food can cause these spiders to cannibalise each other.

Nephila eat flying insects. They spin webs to catch their prey. These are amongst the largest spiders webs in the world, usually being about 1 metre in diameter but can be up to 6m high and 2m wide. The web is based on a Y frame with about 1500 junction points and containing 10-30m of silk. The webs are yellow coloured giving rise to the name 'golden orb' spider. The spider is also known as a 'writing' spider as some of the web appears to have zigzags and other patterns in their design. The yellow colour has two functions: to attract insects and as a camouflage for the spider.

The webs are known as the webs of steel because of the strength of the silk, which compares with Kevlar. This silk is water-based with 30-40% protein polymers. The spiders squirt silk gel from its appellate gland; this undergoes an irreversible reaction when the spider uses its hind legs, body weight and gravity to elongate the gel into a silk thread. This is also a characteristic of liquid crystal materials. The alignment of the protein molecules gives the silk it's great strength. The web substance is being studied for technological development of lightweight, strong materials for the space industry, bullet proof vests, parachutes, seat belts, lightweight clothing and ropes, and artificial tendons and ligaments.

Golden orb spider webs often have complex three-dimensional

structures, such as several horizontal layers, angulations between trees and bushes to cross the flight paths of the flying insects which are their prey. Sometimes birds get tangled in the webs, and although normally spiders leave them, there have been cases of golden orb spiders eating birds.

Webs take about twenty minutes to complete. Unlike other spiders, nephila do not remove and rebuild damaged webs, they repair them. So their webs often look untidy. The spiders hang old carcasses of prey and organic material on their webs; it is thought that this is to warn birds about the web.

In cold windy weather, the golden orb spiders will sometimes take down the bottom of the web, so it is not damaged by the wind. Golden orb spiders are eaten by birds and will jump off the web on a thread. This particular thread is the strongest silk, the silk in the webs having a slightly different property (spiders can produce different types of silk for different purposes). Nephila's web silk is a lot less sticky than other spider's silk.

KEY FEATURES
- Repairs nervous system damage – physical injury to spine (including epidural injections in the spine, trauma from accidents or operations)
- Post operation remedy
- Mends tears in the etheric body and re-establishes boundaries
- Realigns spine and skeletal structure
- Removes blockages in chakras and realigns subtle bodies
- Stomach or intestinal problems, malabsorption syndromes, malnourished states
- Liver and gall bladder disease
- Problems relating to the use of computers or digitalised communication technology
- Reproductive system disorders where the female is dominant and overpowering, or the male is secretive and subversive
- Fear
- Untidiness

PHYSICAL EFFECTS

(Most physical effects are better for motion and warmth, worse for cold and wind)

- Cramping headache – head feels heavy, or skull feels as though it is crushing the brain, that there is a feeling of a band around the head or that the front and sides of the head are tight and restricted
- Excess saliva – needs to swallow a lot
- Removes blockages in upper nose
- Distorted sense of smell – usually pleasant smells are unpleasant and there can be a sense of a pungent odour coming up from stomach
- Pain in brain stem area, tension in back of neck
- Misalignment of atlas or spinal column
- Nervous system damage particularly after epidural injections, operations, injections or trauma to spine from accidents
- Post operation remedy
- Chronic lower back pain
- Limb stiffness
- Paralysis and numbness
- Hand ligaments twisted or in a spasm
- Tension or numbness in feet
- Calcium leaching out of bones particularly associated with digestive malabsorption
- Stomach feels heavy and weighs you down
- Anorexia nervosa
- Vomiting and retching
- Rickets
- Malnutrition
- Mastitis – sense of milk being poisoned
- Constriction in chest – often there are symptoms of heart problems without any physical cause
- Weakness of pericardium
- Liver and gall bladder disease
- Stress with adrenal weakness
- Cramping symptoms
- Fluid retention
- Excessive dryness

- Reproductive problems where the female is dominant and overpowering or has little interest in sex, the male may be secretive and subversive
- Gangrene

MENTAL AND EMOTIONAL EFFECT

- Rigidified stuck thought patterns – can be very dark and depressing
- Fear – either fear of known things or phobias or is caught up by fear based media, fear of getting old, eating wrong things, getting particular diseases etc
- Anger and rage – also the fear underneath the anger and rage
- Inflamed by rage; red hot heart fire blazing, a desire to lash out and destroy what makes you angry, "if looks could kill", people concerned with others giving them dirty looks or a watching them, this makes them angry
- Detached people who want to connect with others, but can't seem to do so, are distanced and divorced from reality either by being in a dream-like state or by having disorganised and chaotic thinking patterns
- Untidiness or disorder
- Feeling of aloneness or separation, cut off from people, or material reality
- People who need to travel or move home frequently to find themselves
- A person who can wait for a long time to act or get something they want
- Females who can be dominant and overpowering or who are so immersed in spiritual thinking or dream-like states that they cannot function in the world
- Males generally are hen-pecked or can be secretive and subversive, can have problems functioning in the world
- Co-dependency in relationships
- People who spend a lot of time on computers and get lost in the Internet (gaming, dating, social network sites etc)
- People who don't have proper boundaries and are emotionally open and exposed so people can take advantage of them
- Feelings of having 'lost' part of themselves or of being invaded by other people or other people's thoughts

- There can be a sense of vulnerability of being trapped or wanting to hold onto relationships and keep people around even when they know that the relationship has no future
- The remedy ultimately grounds you, helps you to connect with others and live successfully on the earth, to connect with one's higher self and spirit
- It brings clarity to ones thought process and transition through difficult emotional states, releasing negativity and bringing in joy
- The remedy helps re-establish proper boundaries between people
- It can bring grit and determination
- Useful to help people suffering from stress and panic attacks, particularly where there is lower back pain or adrenal cramping

EFFECTS ON SUBTLE BODIES / CHAKRAS

- The remedy helps align all the subtle bodies, to mend tears in the aura and establish boundaries so that there are no energy leakages or undue sacrifice to others
- Balancing and promoting flow in the astral body, easing cramping
- Opens up the heart chakra, clearing blockages and enabling healthy heart connections
- Calms down 'heart fire blazing' pathology, e.g. tachycardic arrhythmias, pounding palpitations
- Aligns the chakras
- Helps balance the energy within the antakharana or spinal energy of alignment, with its electrical, gravitational and magnetic waves
- Heals solar plexus chakra, clearing out Atlantean karma and power struggles
- Energies gall bladder meridian
- Opens up crown chakra to enable connections with cosmic thought forms and impulses
- Brings 'light' into the physical body, particularly into the skeleton
- Allows communication between the different dimensions of reality, particularly the 4th and 7th dimensions
- Nephila aids the vital energy and anabolic building-up etheric forces in the body, which aids repair of the nervous system and body tissues

SPIRITUAL ASPECTS

- Nephila provers have had experiences of shamanic dismemberment during the provings, imagining 'their body being torn apart' or eaten – leaving just a dried out skeleton, which then becomes part of an electrified web of light and cosmic energy. Their bodies then undergoing a rebuilding and realignment to cosmic energies and reconnecting to their soul groups and to 'all that is'.
- There have also been strong connections with particular stars, which are relevant to the person concerned. On occasions the map of stars inside the skull vaults has been seen with particular star patterns highlighted.
- Other provers have experienced spiders removing implants and mending tears in their etheric body. Some provers experienced 'soul retrieval', with lost soul parts (fragmented off due to karma) being reunited with them. This remedy is useful if the client is having shamanic soul retrieval.
- Nephila, as in other spider remedies, is good for people trapped in duality based, digital computer technology. This also involves the Lucifieric forces of bringing in the light and shining it into the dark places in ones life, helping people to remember the interconnectedness and multidimensional unity based nature of reality and link back to their origins in the stars.
- The remedy helps repair the nervous system at a physical level and promotes spiritual impulses and connections from different planes of reality, higher spiritual beings and cosmos. It helps the brain function as a receiver and transmitter of spiritual thought and helps one to 'manifest their own reality' by drawing events and people towards them.
- Nephila can help meditation by putting you in a calm, quiet state, open and reconnected to the spiritual world.
- Nephila aids communication between people and the link into one's soul group. This can help bring up issues such as co-dependency and karmic ties, to reach a place where there is a fostering of respect for others and their own individual journey through life. It may bring to realisation lost soul parts, or other people caught within your energy field, particularly in your heart which feels heavy and burdened with a sense of indistinct boundaries, other people then

taking advantage of this. This sometimes leads to a state of rage caused by invasion of personal space.

- There may also have been many house or job moves or a lot of travelling 'to find myself' .
- The remedy helps to bring a sense of being at home in one's heart, wherever in the world you happen to be, and brings a sense of empowerment, liberation and freedom, of oneness and integration of all aspects of you being. It also brings a connection between the personality and soul, of the force of grace from the seventh dimension of reality.
- It helps remove implants, aids soul retrieval, and cuts karmic cords and power struggles. It can help connect to stars and planets, help connect to the spiritual world giving divine guidance about one's purpose and path through life and helping define current aims and objectives.

OTHER

- Feeling of light flowing through the skeleton
- Feeling of being plugged into a grid
- Feeling of a steel rod in the spine
- Experiencing a heaviness of physical body particularly around the stomach, feeling held down and trapped by it
- Aids sporting and sexual performance
- If given to both parties in a relationship it can help them to link their energies and bring in higher spiritual aspects to their chi
- Brings order to chaos

RELATED REMEDIES

- Alumina – rigidified, dried out with degeneration of the mind
- Ant crud – helping synchronise the subtle bodies and brings order to the nervous system
- Calc phos – bringing in light to the structure of the bones
- Emerald – energetic connections with other beings through the heart centre
- Kali remedies – especially Kali bich, healing the etheric body and opening up the heart to unconditional love

- Phosphorus – bringing light into the subtle bodies
- This remedy has been used successfully with people who have been traumatised by war and violence either in this lifetime or in past lives

APPENDIX

THE DIFFERENCE BETWEEN BUGS AND BEETLES

Although initially it may seem difficult to distinguish between the Bugs (Hemiptera) and the Beetles (Coleoptera) their outward appearances and the nature of their lifestyles points to them being at very different levels of the insect world. The cockroaches (Blattodea) also are different types of insects which sometimes are wrongly thought of as being bugs.

Bugs essentially have an outward appearance of folding their wings flat over their bodies in a characteristic X shape, whereas Beetles have closed wings which meet in a straight line down their backs.

Bugs have forewings which are leathery and thickened at the front but membranous at the back (hence the name Hemiptera – 'half wing'). Beetles have leathery front wings with an electra to cover and protect their fragile, membranous, flying wings.

Bugs have mouthparts adapted to piercing and sucking with straw-like rostrums, as they feed mainly on plant sap or blood. Beetles have specialised mouth parts which are well adapted to their individual diets e.g. some carnivores have sharp sickle-shaped jaws, some plant feeders have snouts with biting jaws at the top. Most beetles have 'biting' mouth parts.

The life cycles of bugs are characterised by 'incomplete metamorphosis' similar to the Orthoptera (grasshoppers, locusts, crickets) where they lay eggs, the eggs hatch to become nymphs which often resemble the adult insects. They grow and undergo a series of moults resulting in them increasing in size and growing progressively larger wing stubs. The final moult leaves them in the adult form with wings which are able to fly. The nymphs feed on similar things to the adults and sometimes share the same habitats.

The life cycles of beetles show 'total metamorphosis', as do the Lepidoptera (butterflies and moths). They lay eggs which hatch into nymphs or larvae. These often are very different from the adults of the species, eating different foodstuffs (having adapted biting mouth parts) and sometimes living in completely different habitats e.g. the Long

Horned Beetle larva eats wood from inside rotting tree stumps, whist the adult feeds on nectar and pollen from flowers. The Hercules beetle starts life as a small larva eating plant leaves, the adult is a very large beetle well adapted to life on the rainforest floors. The 'immature' forms, like the bugs, (which are effectively 'eating machines') undergo a series of moults where they shed their skins and grow larger. However, they then pupate, spinning a pupa around the immature form, and within the pupa there is a complete breakdown and transformation of the form of the beetle down to a cellular level. A completely new adult form is generated, which is very different from the immature insect. The adult emerges from the pupa with an urge to reproduce and fly.

It is often very difficult to tell bug nymphs from beetle larvae whilst in their immature forms. Bug nymphs always have legs, some beetle larvae do, but not many. The underside of bug nymphs has a needle-like rostrum projecting down from the head or folded under the body, whilst beetle larvae have biting mouthparts, often with conspicuous jaws.

THE ESOTERIC NATURE OF BUGS

Essentially bugs are of the Order Hemiptera (Bugs) and range from minute wingless insects to giant water bugs, large enough to feed on frogs and small fish. Like beetles they are found in all terrestrial habit tats, in fresh water and on the surface of the northern oceans. They include predacious, 'blood sucking' species e.g. lice, carnivores and herbivores. Most plant feeding bugs contain symbiotic bacteria in their digestive tracts.

The bugs are further classified into two sub orders: The Heteroptera (True Bugs) and the Homoptera (including tree and leaf hoppers, cicadas, lice, aphids and scale insects).

Homoptera are bugs where in the adult form they have membranous wings held out at a 'tent like' angle above the body, the wings are uniform in texture, rather than being divided into leathery ad membranous areas. Homopterans often give off sweet smelling secretions similar to honey dew. Heteroptera (True Bugs) are characterised by an X shape made of two triangles on their backs (the shield bug form is typical of this). The forewings are held flat against the body so the membranous areas of the wing tips overlap to form the first triangle, the second triangle being formed by the scutellum.

The formation of the wing pattern on the back of bugs shows a more advanced formation than that on the back of beetles according to the Laws of Synchronicity. The beetles essentially are split left to right, ie. a horizontal split (which leads to issues of directionality, e.g dyslexia, dyspraxia etc, but essentially keep the beetles on the horizontal plane and the duality based 'material' world. The triangular shapes on the back of bugs allow for a vertical shift, so they are more able to lift themselves up from the earth plane and connect with the spiritual realm. Triangles also mean that duality can be overcome as they allow for different connections with other dimensions. This wing formation appears to be the direct opposite of where the bugs and beetles stand in terms of their life cycles, with the beetles undergoing total metamorphosis, this raises them up to be able to escape from the earth plane. However, bugs are working their way away from the earth to solar energies, whereas beetles have already attained this but have decided to return to Earth for their service, to shine the light of spirit that they have attained on the different kingdoms to help evolutionary advancement. The vertical split on their backs provides them with a hold onto the earth.

The triangles on the back of bugs represent the trinity in all its aspects.

The compound eyes of bugs and beetles enable them to 'view' the whole of the environment all the time in a soft focus so they do not see it as separate from themselves and feel their interconnectedness with all things. This is different from human consciousness, where our eyes view individual things in sharp stereoscopic focus and this leads to feeling that we are separate from our environment rather than being an interconnected part of the whole. The 'interconnectedness' of the insect world is further developed because they have group souls, rather than the individual souls of humans. This enables bugs, and other insects to communicate with different dimensions and different levels of consciousness. They are particularly connected to the fifth dimension of reality with its quality of mental or thought-form structure, but due to the 'incomplete' metamorphosis are less able to adapt quickly to changes in environmental conditions than beetles.

Bugs are found in a huge range of habitats. Some live their entire lives on the forest floors, eating dead plant and animal material and recycling this to serve both plants which use it to grow in, and recycling it back to the mineral kingdom as soil. They live their lives in the darkness or dimly lit conditions of the rainforest floor but aspire to fly up and leave the realm of the earth for the realm of the air. Bugs are linked to the darkness of the earth and also to lunar forces. Therefore they are working with the past, clearing old memories, both ancestral and those of their present lifetimes. They work with clearing out old unwanted energies and balancing the etheric and astral forces on the forest floors and in the earth. So they are useful to help clear out entities, work with ahrimanic duality based energies, possession etc.

Some bugs live in water, where they are separate from the earth. They have thin legs, the evolution of their limbs being related to the involution of the lungs. Bugs living in the water are able to do so by trapping water between their chitinous bodies and electras.

Other bugs live higher up on plants and are higher up the evolutionary scale in raising themselves up above the earth plane, so they deal with bringing in future codes and deal with the astrality around the plant leaves, however they are limited by not undergoing total metamorphosis.

Symbiotic relationships, particularly with bacteria within their guts, characterise bugs. In the insect / plant relationship within a food chain a mutually agreeable 'love based' dialogue takes place where both get what they need rather than simply being a matter of survival and nutrition.

A key aspect to bugs is their special adaptation to 'sucking' in liquid food, either from plant sap, nectar or blood (e.g. reduvid bugs). The hypopharynx, which is small and insignificant in creatures with biting jaws, like beetles, is important because it is the pumping chamber which draws up the liquid food. Sucking then utilises and combines the etheric energies of the insect with that of the food source, promoting tissue growth, nutrition and metabolism in a mutually beneficial way. Tubes are ruled by astral forces which cause movement of the fluids and help lead the air element through the fluid.

Most bugs can fly. The movement of all flying insects is permeated by 'lofty wisdom' of the group soul. Flight comes from the ability of the subtle bodies to respond to this wisdom, which results in movement in the astral plane and movement of the physical wings of bugs. This ability attracts the agni fire spirits to the bugs and when they fly they are often accompanied by fire spirits flying in the solar realm.

Many bugs have varied antennae, which pick up signals from their surroundings e.g. some can pick up pheromone signals from up to a mile away.

www.ingramcontent.com/pod-product-compliance
Lightning Source LLC
Chambersburg PA
CBHW022111280326
41933CB00007B/342